SCIENCE AS SALVATION

SCIENCE AS SALVATION

A modern myth and its meaning

Mary Midgley

Routledge
Taylor & Francis Group

LONDON AND NEW YORK

First published 1992
by Routledge
2 Park Square, Milton Park, Abingdon, Oxon, OX14 4RN
29th West 35th Street, New York, NY 10001

Paperback edition 1994, 1996

Transferred to Digital Printing 2004

Typeset in Palatino by
Falcon Typographic Art Ltd, Edinburgh

British Library Cataloguing in Publication Data
Midgley, Mary
Science as salvation: a modern myth and its meaning.
I. Title
501

Library of Congress Cataloguing in Publication Data
Midgley, Mary
Science as salvation: a modern myth and its meaning/Mary
Midgley.
p.   cm.
Includes bibliographical references and index.
1. Science – Philosophy, 2. Religion and science – 1946–
I. Title.
Q175.M613   1992
501 – dc20   91–30984

ISBN 0–415–06271–3 (hbk)
ISBN 0–415–10773–3 (pbk)

Printed and bound by Antony Rowe Ltd, Eastbourne

For Davy

CONTENTS

CONTENTS

ACKNOWLEDGEMENTS

I am most grateful to the University of Edinburgh, who invited me to deliver, in the spring of 1990, the series of Gifford Lectures from which this book grew, under the title 'Science and Salvation'. I had an enormous amount of help there, both from the kindness of the Administration and from the lively and sympathetic audience which so surprisingly grasped my rather puzzling aim, and generously encouraged me to struggle on with a vast and difficult subject. Lord Gifford himself, too, seems to me to deserve a great deal of thanks for inventing this admirable institution.

Since then, I have had all sorts of help from a great many people, and have indeed continually bothered everyone I knew who had a scientific or theological training with questions. People attending many conferences, notably those of the Forum for Science and Religion, have been most patient and responsive about this. (I would name particularly Peter Hodgson, Lawrence Osborn and Christopher Finlay.) So have my friends, colleagues and family in Newcastle. Dr Ursula Philipp gave me invaluable help in understanding the ideas of J. B. S. Haldane, who played a crucial role in this story. Ian Ground and David Midgley have been enormously helpful in finding suitable scientific sources, explaining the background and sorting out the conceptual tangles that it raises. Finally, Professors Christopher Isham and Ivan Tolstoy have both very kindly read two longish sections of the book to check for scientific and other errors, and have made extremely helpful comments. Any blunders that remain are, of course, entirely my own.

ACKNOWLEDGEMENTS

Note on Notes: The scope of possible reading on this vast subject seems to me so endless that any attempt at a full bibliography would merely be hopelessly unwieldy and confusing. Except in a few chapters where the history is important, I have therefore given references only to relatively few books, most of which are not difficult to read. These books themselves contain further references, which will I hope enable readers to follow matters further. Endless controversies, which I have summed up here with brutal brevity, can of course be traced out at their proper length elsewhere. But I think that all of us, even the most learned, need at a times a more inclusive view, and that is what I have aimed at here.

1

SALVATION AND THE ACADEMICS

SALVATION IS NO LONGER OFFICIAL

The idea that we can reach salvation through science is ancient and powerful. It is by no means nonsense, but it lies at present in a good deal of confusion. Its many strands – some helpful, some not – greatly need sorting. In the seventeenth century, when modern science first arose, it was an entirely natural thought. The great thinkers of that time took it for granted as central to their endeavour. Nature was God's creation, and to study it was simply one of the many ways to celebrate his glory. That celebration was understood to be the proper destiny of the soul, the meaning of human life.

Since that time things have changed greatly. For a number of reasons, God has been pushed into the background. The conceptual maps that he once dominated go on, however, being used as if they did not need much revision. This makes trouble on many issues, and notions about the special saving power of science are among them.

Does this language of salvation seem alarmingly strong? I use it because I want to stress throughout this book how deeply these matters affect all of us, not only scientists and not only intellectuals. Any system of thought playing the huge part that science now plays in our lives must also shape our guiding myths and colour our imaginations profoundly. It is not just a useful tool. It is also a pattern that we follow at a deep level in trying to meet our imaginative needs.

This book is therefore not just about our attitudes to science but about those imaginative needs. It is about myth-making, not just as a private vice, but as a vital human function. The way we use science for this function is, however, today not an acknowledged academic topic. Officially speaking, academic studies don't now offer salvation at all. Their journals certainly don't expect to be used by people desperately seeking for the meaning of life, and such people could usually not read them anyway. As in the Tower of Babel, each discipline speaks only in its own tongue. There is no interdisciplinary language for discussing the relations of studies to one another, nor to the world around them. Least of all is there any such language for considering the general meaning for us of each study, the part that it plays in life.

People who are rash enough to discuss these things must, then, use ordinary speech. However carefully they think, they tend to be classed as informal operators, expressing merely 'intuitions' (a name recently invented for views not officially stamped by any university department). This deliberate self-isolation is specially marked in the physical sciences, where it is often fatalistically supposed that serious work cannot be explained at all to outsiders. Yet there are bold and clear-headed explainers who do manage to do that hard thing.[1] This work is surely of the first importance, since intellectual enquiries, like nation-states, always do have outside relations which can matter greatly to them. They all draw concepts, presuppositions and metaphors from outside their borders, items which can deeply affect their inner working.

HOW SCIENTIFIC ARE MOST OF US?

Obviously, this increasing technicality in the sciences has served very important functions. What makes it troublesome today is that it leaves unserved the general need for understanding, and whatever spiritual needs lie behind it. The promise of satisfying those spiritual needs has played a great part in establishing the special glory of the abstraction 'science' in our culture, and in forging the idea that we are a scientifically-minded people. It has built up a strong emotive and romantic conception of 'science' as a spiritual power – a most ambitious estimate of what this abstraction is and can do.

The retreat of the specialists has not wiped out this estimate. It is normal thinking today that we – all of us – not only depend practically on applied science for our lifestyle, but also have, and ought to have, our general thinking shaped by pure science – by its theories. That we are 'scientific' in our attitudes and live in a scientific age is widely held to be both a fact and a ground for rejoicing, an achievement to be celebrated and carried further.

I am not, at the moment, complaining at all of this wish and this celebration. I am asking what they mean. As the gap between professional science and everyday thinking widens, it gets increasingly hard to work out in what sense most of us can be said to be thinking scientifically at all. What science we do know we know at second-hand, on authority about which we are usually vague. It is mostly not up-to-date. Is this second-hand and out-of-date science enough?

At present it is common to reply that it is indeed enough. Many scientists will now say flatly that most of us cannot expect to understand what is happening at all, and had better not even mess around with the popularizations. This gloomy estimate must extend, of course, far beyond the uneducated proles to the scientists themselves, when they deal with anything outside their own increasingly narrow provinces. There cannot, on this view, ever be such a thing as a scientifically-minded public.

Yet the idea still persists that science is not just certain people's trade, but a universally important ideal. We tend to believe that it is the duty and hope of us all to be in some way scientific, and this is certainly not seen just as a matter of practical convenience. Science is seen as having a special kind of value to which we all owe allegiance. People who want to list the glories of our civilization are almost sure to list science – meaning primarily physical science – among them, along with art. And the special value of science, like that of art, is not supposed to reach only the few who produce it, but also the public which receives it.

Recent worries about the dangers that may flow from technology have not really changed this way of thinking. These dangers are still mostly attributed to misuse of science rather than to science itself. There has indeed long been an explicit anti-science strain in our culture, with impressive ancestors such as Blake, and it has gained some strength lately. But it is still a descant; the main anthem is still one of praise. And until the last few decades, many acute and polymathic scientists were happy to explain why this high estimate of their occupation was justified.

3

FROM TOO MUCH TO TOO LITTLE

They are far less willing to do that now. Many scientists, if they are asked what – beyond its obvious usefulness – is the function of science, will either evade the question, make vague euphoric noises, or give answers that seem almost pathologically modest, parsimonious and negative. They claim that they are merely humble standard operatives in an immense, impersonal falsification factory, busied solely in examining an endless succession of detailed hypotheses about the physical world and in proving most of them to be false, by a single, prefabricated 'scientific method'. In slightly less stern and more realistic moods, they may mention a conceptual background out of which these suggestions arise. But unless they have had some historical training, these scientists are most unlikely to suggest that this background could have anything to do with the rest of human thought, still less with the rest of life. The isolation of 'science' from other topics is widely held to be necessary for its purity.

There has been a remarkable move from claiming far too much to claiming far too little. C.H. Waddington, in his book *The Scientific Attitude* (1941), noted what had already begun to happen:

> Responsible scientists, looking at their colleagues, saw the obvious fact that most specialists were quite unfitted to play an important part in the evolution of general culture; but, far from acknowledging that this was a sign of science's failure, they accepted it almost with glee as an excuse which let them out of the necessity of thinking about wider issues.[2]

In fairness, we should notice that many specialists in the humanities do this too, and with even less excuse. But the special hopes that the age places on science make its withdrawal a particularly serious matter. Nobody today supposes that the distinction of our epoch depends on its being a historical, or a literary, or a philosophical one. But they do suppose that about its being scientific.

It is easy to see how the specialists' rather frantic modesty arose. It was a reaction against excess. Philosophers of science invented it as a way of disqualifying the Marxist and

4

Freudian sages who claimed the prestige of science for their vast metaphysical systems. In particular, in Britain, the last great generation of Marxist polymaths – Needham, Bernal, Haldane and the rest – were most alarming to less well-educated scientists. They were charismatic, popular and learned writers, using the authority of science skilfully to back their political views. Thus, Bernal argued that communism was simply the logical conclusion of the whole scientific endeavour:

> Already we have in the practice of science the prototype for all common human action . . . The methods by which this task is attempted, however imperfectly they are realised, are the methods by which humanity is most likely to secure its own future. In its endeavour, science is communism.[3]

Now people like Bernal could certainly have been answered. But answering them was not specialized scientific work; it involved wider thinking. More orthodox scientists who wanted to avoid this saw that it would be easier to outlaw these unfairly well-educated sages instead by narrowing the idea of 'science', so as to shut their kind of speculation out of it by definition.

They therefore contracted science and pulled up the draw-bridge. A disturbance followed when it was noticed that they had accidentally left the whole of evolutionary theory outside in the unscientific badlands as well. But special arrangements were made to pull it in without compromising the principle. That principle was to minimize the business of 'science' – to define it as narrowly as possible, confining its prestige to detailed, provable, specialized work.

THE EXALTATION OF NOT BEING WRONG

Was this wise? People choosing this policy were assuming that the prestige, the value of science centres on never making a mistake – on precision, specialization and infallible correctness. But is that its real point? Science surely has a more positive value, both for the world at large, and for scientists themselves when they are not making this kind of defence. The glory of science is not that it never makes mistakes, which is plainly false anyway. It is much more a matter of dealing with supremely interesting topics – matters that can seriously affect the way we see human life.

For instance, the conception of order in the universe is a crucial background to our thought, and just how that order is conceived – just what kind of causal necessity we picture – affects the whole arrangement of our concepts. Again, the way in which we think of the relation between our own species and all other living things is an essential element in shaping our inner maps. So is the notion that we have of 'life' – of the meaning of the difference between what is living and what is not. *Aristotle*

Moral Law

The idea that science really matters, that it has a key place in shaping the rest of thought, still prevails, and is far more than just a conviction of its indirect usefulness through technology. When Karl Popper – often inclined to minimalism – made in 1972 the startling claim that science is 'perhaps the most powerful tool for biological adaptation which has ever emerged in the course of organic evolution',[4] he was plainly thinking of it as something immensely larger than the accumulation of unconnected, detailed, negative facts. He was indeed claiming for it a status considerably higher than the one Waddington had outlined for his own much wider conception of science thirty years earlier. Waddington said:

> Science *by itself* is able to provide mankind with a way of life which is, firstly, self-consistent and harmonious, and, secondly, free for the exercise of that objective reason on which our material progress depends. So far as I can see, *the scientific attitude of mind is the only one which is, at the present day, adequate in both these respects.* There are many other worthy ideals which might supplement it, but I cannot see that any of them could take its place as the basis of a progressing and rich society.[5] (Emphases mine)

This exalted status obviously could not be claimed for a mere batch of stored facts, however large. Stored facts are like stored tools or stored musical instruments, valueless unless you know how to use them, how to connect them with other things, how to understand them. It is surely the interpretative scheme, not the stored data waiting to be interpreted, that we have in mind when we make large claims like this for the value of science.

6

THE HIGH HOPES

Clearly, many people still think science quite as important as
Popper and Waddington did. Putting it crudely, many people
have long looked, and still do look, to science for an important
aspect of their salvation, and these are by no means only
people who themselves know much science. If the public had
not to some extent shared this hope, it would scarcely have
spent even as much money as it has on pure research. Many
scientists themselves, too, would probably not have chosen it
as the occupation of their lives if they had not agreed with
them. There are many branches of science, perhaps particularly
in theoretical physics, which students choose because of a vision
of how the world fundamentally is, a vision in which they have
faith and which they want to follow out in detail.

In spite of today's official modesty, large claims revealing this
kind of faith still constantly appear in books that officially do
nothing to back them. There has, indeed, recently been an
exuberant expansion of claims to moral and intellectual territory
which earlier pioneers of modern science sternly disowned.

In particular, there are today what seem to be renewed
offers of an explanation in terms of purpose – something
which physical science has officially forsworn since the time
of Galileo. Thus, Richard Dawkins joyfully proclaims that, since
we now have modern biology, 'we no longer have to resort
to superstition when faced with *the deep problems; Is there a
meaning to life? What are we for? What is man?*'[6] (Emphasis mine).
Dawkins offers science as able to deal with all that, and as the
only alternative to superstition in doing so. Similarly, Stephen
Hawking speaks of his cosmological enquiry as a response to
an ancient, timeless human longing:

> Ever since the dawn of civilization, people . . . have craved
> an understanding of the underlying order in the world.
> Today we still yearn to know *why we are here* and where
> we came from. Humanity's deepest desire for knowledge
> is justification enough for our continuing quest. And our
> goal is nothing less than a complete description of the
> universe we live in.[7]

A complete description? Is there such a thing? Since there is in
principle no limit to the questions that might need answering,

it is not a clear idea, but does it even point in the right direction? Would a complete description, of the kind that could be approached through science, be the kind of answer expected by the question 'why are we here?' Hawking writes that, when a satisfactory cosmological theory has emerged,

Even if we did find such evidence, would people be content to accept after centuries of religious faith?

> we shall all – philosophers, scientists and just ordinary people – be able to take part in the discussion of the question *why it is that we and the universe exist.* If we find the answer to that, it would be the ultimate triumph of human reason – for then we would know *the mind of God.*[8] (Emphases mine)

Old ideas vs. New ideas

In what sense could modern cosmology be pointing towards that? Does it seem plausible that this ancient, universal human longing was always a desire for the kind of scientific theory that Hawking and his colleagues now hope to forge? This would be strange, since before the last few hundred years, nobody anywhere ever dreamt of looking for that kind of theory. Even today very few people in the world have heard of it.

The ancient desire was surely a quite different one. It was a desire for kinds of explanation that are both much wider and more immediate. The wish to know 'why we are here', unmistakably asks a question about the point and purpose of existence. The word is 'why', not 'how'. The ancient question is not about the remote physical causes that may have made that existence possible; it is a purpose-question; it is teleological. The phrase, 'the mind of God', too, could scarcely cover a mere account of causes. It cannot avoid referring to purpose.

Define God

Incidentally, the word God, which suddenly appears at several points in Hawking's argument, badly needs explaining. It is notoriously a most obscure and ambiguous word, yet it gets no discussion and does not figure in either the index or the glossary to *A Brief History of Time.* It is treated as unproblematic. Hawking doesn't, in fact, seem to have heard that many people – anthropologists, historians of thought, philosophers, theologians – have already done a lot of useful work on such matters, have detected many of the more obvious bugs in the program, and could have saved him some unnecessary confusion.

8

QUESTIONS ABOUT TELEOLOGY

[handwritten annotation: Reasoning from Purpose]

This is not just a cheap jibe at Hawking. The point is central to our theme. Teleology – reasoning from purpose – is, I believe, a much more pervasive, much less dispensable element in human thought than has usually been noticed. I will suggest that it is doubtful, in fact, whether our imaginations can work at all without it. General attacks on it have often indeed exposed misuses of it – pieces of bad and ill-controlled teleology. But the idea of dropping it altogether may not be much more practical than that of stopping breathing. Purpose-centred thinking is woven into all our serious attempts to understand anything, and above all into those of science.

What this large and perhaps alarming suggestion means will, I hope, gradually become clear. Briefly, however: Understanding anything is finding order in it, and, for human thought, the idea of order seems necessarily to carry a background context of planning, of intention. Obviously, this is primarily a remark *[handwritten annotation: Humans require order.]* about what our minds demand – about the ways of thinking possible to us – rather than directly about the universe. But *[handwritten annotation: Religion provides order.]* then our minds are what we have to use, and we need to be aware of their workings.

The connexion between order and planning comes out in the range of words we use to describe order. *Order* itself and *direction* both also mean *command*. *Design, system, arrangement, construction, structure, formation, plan, scheme, law, rule, program, mechanism* and *organization* all mean some kind of intentional composition. *Pattern* turns out (rather surprisingly) to be the same word as *patron*, meaning source or authority.

And so on. The recent adoption of information-language is just one more very striking step in that bold process of assuming the penetration of mind through matter that has made Western science possible. The Greek word *cosmos* (akin to *cosmetic*) simply meant *arrangement* or *adornment*. The Judaeo-Christian concept of purposive organization by a single Creator reinforced this confident approach, which was reformulated in the seventeenth-century use of Plato's idea that God was the Great Geometer. When we carry this policy still further today, we only differ from our predecessors in being curiously unwilling to notice what we are doing.

The vocabulary just mentioned is not (I am suggesting) some

9

irrelevant superstitious survival. These words are indeed metaphors. But they are not optional, disposable metaphors. They cannot be replaced at will by literal and 'objective' language. Like many metaphors, these form part of the thought. As with the physical terms that we use to describe mental processes – seeing, grasping, missing, clarifying, obscuring – these are the most direct words available.[9] Because we consider order as something readable by our minds, we have to think of it as a communication, as meaningful. But meaning unavoidably strikes us as an expression of mind, not as something alien to it.

TELEOLOGY AND TIME

Using this mental category does not have to involve the means–end pattern through time which, for some reason, people tend to think is the whole of teleology. It need not involve planning done at a certain time to produce a result expected later. That 'consequentialist' or jam-tomorrow pattern is in fact quite a limited part of it. Aristotle, who first analysed the different sorts of functional reasoning, strongly noted its inadequacy for describing human purposes. The best and most central human activities are (he said) actions done for their own sake, done because they have value in themselves. Means-to-end calculations are subsidiary, they are plans devised to make these self-rewarding activities possible.[10]

For instance, neither thinking nor singing nor talking to one's friends need be done as a means to something later. Again, the first notes of a song are not a means to its cadence, nor the first ten years of a friendship a means to its final end. The essential teleological question is not 'what later thing is this leading to?' It is, more widely, 'what is this for? what is the point of it? what part does it play in a wider whole?' Acts like singing are intentional – they are done 'on purpose' – but not for the sake of producing consequences.[11] The essential relation involved is not that of earlier to later time. It is that of part to whole.

This point about the broad scope of teleological reasoning needs noting at once. It affects our theme in two ways. Morally, it is relevant because the startling plans for human immortality that we shall shortly be considering are an extreme example of consequentialism – a profound shifting of moral attention away from present problems to incalculably distant future jam. And in physical science, the working of this timeless yet functional thinking helps to explain the relation of teleology to our thoughts about inanimate things.

Understanding the pattern in a crystal or a river system is not discovering what somebody once designed it to produce. It is simply putting it into the class of things meaningful – noting how its parts relate to it as a whole, and how it itself relates to the larger scene around it. It is reading it. But that – for us – does involve understanding it in the way we understand a communication. The responses we make to it, the faculties by which we deal with it, are unavoidably those by which we would take in social messages. It falls into the department of mind.

WHAT'S IN A NAME?

A brief word may be needed here about the term *teleonomy*. This was invented by nervous biologists to replace the word *teleology* by describing functional behaviour in organisms without (as they hoped) implying the presence of a designer.[12] This move, however, quite underestimates both the traditional scope of the word and the underlying problems. 'Teleological' is the name of a kind of explanation, namely, one that works by mentioning a function – not, for instance, by mentioning a cause. (A most troublesome traditional mistranslation of Aristotle is at work here, producing the idea that he thought of purposes as a kind of causes – 'final causes'. In fact, what are called his 'four causes' are four sorts of explanation, and this is simply the one that answers the question 'what for?')[13] All talk of function is therefore in any case teleological. It is about design. What relation this fact may have to the possible presence of a designer is a separate question. People hoping to settle the whole issue by using the word teleonomy commonly take this further question to be finally settled by naming either Darwinian natural selection or (better still) blind chance as the quasi-designer. We will discuss these solutions later.

HOW MUCH DOES MEANING MEAN?

The idea that we need to think teleologically is not fashionable today, and may be dismissed as extravagant. I will suggest that that dismissal grows less plausible once you notice the extravagance and implausibility of the views that are supposed to displace it – the bugs now infesting the idea of radically mindless matter. The suspicious reader can perhaps put off worrying about my suggestions for curing these infestations until we have had a good look at them.

It may be best, too, to repeat that there is no need at the moment for that alarmed reader to reach for the anti-God button. This attribution of meaning to orderly phenomena is something extremely modest, mild and general. It is nothing as bold and specific as an Argument from Design to an intelligent, humanoid creator. It is simply the assigning of orderly things by our minds to a different mental category from the 'buzzing, blooming confusion' that wallows behind them, the disorderly background of undigested experience.

The point is just that this category of the intelligible necessarily counts as akin to mind, because the order we detect in it is of the kind our minds acknowledge. It is quite true that the religions have grown out of this unifying, ordering vision. But then, so have the sciences. The kinship between these two ways of thinking is far closer than has been recognized. The idea that being scientific simply means being irreligious is a particularly naive one. It has caused a lot of confusion and will get us nowhere.[14]

Anyone who doubts this might like to try the experiment of finding more suitable, antiseptic words to replace the religious language used in a certain famous exchange between Einstein and Bohr. Disturbed by the implication of real disorder in Bohr's interpretation of quantum mechanics, Einstein said, 'God does not play dice'. Bohr replied, 'Einstein, stop telling God what to do.'[15] Because Bohr is held to have won this debate and his views are still widely accepted, this conversation is now widely quoted in discussions of the topic. But those quoting it seldom offer a carefully secular paraphrase to show just what he had established, nor do they explain why this language struck these great men as so well fitted for their purpose.

The close dependence of all scientific explanation on mental concepts has become still clearer lately in the widespread use of terms like *communication* and *information* to describe all sorts of non-conscious interactions. Why are these metaphors proving so helpful, so enormously convenient that some people do not notice they are metaphors at all? Such people innocently suppose that to say 'DNA contains the necessary information' is to say something as straightforward as that it contains the necessary carbon and hydrogen.

More perceptive writers guard themselves against animism by explaining that DNA does not literally think or talk. But they still do not often ask themselves *why* it should be so helpful – indeed so essential – to go on as if it did. Exactly what parts of the comparison are useful? Why is it so necessary? How should we speak if we were not allowed to use it?

The use of such categories is, I believe, a necessary condition of the way our minds work on such subjects. We understand today that it is a bad idea to exterminate the natural fauna of the human gut. But trying to exterminate the natural fauna and flora of the human imagination is perhaps no more sensible. We have a choice of what myths, what visions we will use to help us understand the physical world. We do not have a choice of understanding it without using any myths or visions at all. Again, we have a real choice between becoming aware of these myths and ignoring them. If we ignore them, we travel blindly inside myths and visions which are largely provided by other people. This makes it much harder to know where we are going.

Acknowledging matter as somehow akin to and penetrated by mind is not adding a new, extravagant assumption to our existing thought-system. It is becoming aware of something we are doing already. The humbug of pretending that we could carry on intellectual life in an intrinsically unintelligible world is akin to the humbug of pretending that we could live without depending on other people. Just as we wildly claim to stand only on our own feet, without any help from others, so we wildly claim that we would be quite capable of 'imposing order' on an intrinsically disordered universe. In both cases, we take for granted an external support without which we could not live, and pride ourselves on managing so cleverly without it. There is nothing parsimonious about this kind of conceit.

We need others

WHY DOES SCIENCE WORK?

Behind these questions lies a vast issue about mind and matter, which this book cannot of course resolve. I am trying here only to get past a few bad supposed solutions to it, which at present block thought on a really interesting topic. As may be plain, this topic is essentially the one which caused Einstein often to remark that the really surprising thing about science is that it works at all. Puzzlement does not arise out of some eccentric and optional religious enquiry, but out of the simple observation that the laws of thought turn out to be the laws of things. As C.S. Lewis put it:

> We find that matter always obeys the same laws which our logic obeys . . . No one can suppose that this can be due to a happy coincidence. A great many people think that it is due to the fact that Nature produces the mind. But on the assumption that Nature is herself mindless, this provides no explanation. To be the result of a series of mindless events is one thing; to be a kind of plan or true account of the laws according to which these mindless events happen is quite another . . . It is as if cabbages, in addition to resulting from the laws of botany, also gave lectures in that subject . . . We must seek the real explanation elsewhere.
>
> I want to put this other explanation in the broadest possible terms and am anxious that you should not imagine I am trying to prove anything more, or more definite, than I really am . . . *Unless all that we take to be knowledge is an illusion, we must hold that in thinking we are not reading rationality into an irrational universe, but responding to a rationality with which the universe has always been saturated.*[16] (Emphasis mine)

As he notes, this might lead to many sorts of philosophical positions, not necessarily theistic ones. To find our way, I shall simply try in this book to do something crude which is often helpful in such cases – namely, to point out some very bad ideas that are currently accepted. By seeing what not to think, we can often move towards the parts of the map which will help us. Besides, the appeal of certain mistakes often lights up aspects of the problem which we would otherwise miss.

Let us look first at ways in which the supposition that matter is totally alien to mind is now proving incoherent. Like many other people, I shall point out how odd are the notions, both of matter and of mind, which we have inherited from Descartes, and on which their supposed total separateness was originally grounded. I shall then discuss, what has been rather less noticed, how much odder and more unrealistic the idea of matter has grown when attempts are made to amputate mind, leaving it in limbo.

LISTENING TO THE IMAGINATION

In considering all this, I believe that we must attend seriously to myths, metaphors, images and the other half-conscious apparatus of thought surrounding the official doctrines. I shall point out strange compensatory fantasies found in the work of various scientific writers – some of whom have been, in theory, austerely bent on disinfecting the world of traditional teleology – noting how they often seem to end up with a far cruder, less rational teleological doctrine than those they were attacking. Throw purpose out through the door and it seems to creep in up the drains and through the central heating. (I have discussed this matter in an earlier book[17] in relation to terrestrial evolution, but I had not then noticed how cosmologists were developing it on the celestial scale.)

Scientific reviewers, when discussing writings of this kind, often treat the myths as a side-issue. Concentrating on what is acceptable as science, they expect the rest to fade away harmlessly into the general culture. But it does not necessarily do this. It can hang around like a fog, changing the atmosphere of thought and influencing ideas quite strongly. It tends to be the part of a book that people remember. In particular, it can be expected to have a strong effect on students.

Attending to the workings of the scientific imagination is not a soft option, and it is not mere gossip. This material has (as I shall suggest) a far closer, more organic connexion with our official thinking than may appear. It is not just a harmless, licensed amusement. It plays a part in shaping the world-pictures that determine our standards of thought – the standards by which we judge what is possible and plausible.

THE REINFLATION OF PHYSICS

Recent attempts to make traditional materialism consistent have (as we shall see) often resulted in making it romantic, superstitious and irrationalistic. There have also been lately, as already mentioned, a number of attempts by cosmologists to expand materialism by recolonizing, as an official part of their subject, territories formerly ruled illicit, in particular, by relegitimizing teleology.

Since the Big Bang became widely accepted, the urge to find some sort of purposive story for the cosmos has become almost irresistible. (If, as now seems possible, theorists dissolve the big bang again, it will be interesting to see what becomes of this trend.) This attempt to think about cosmic purpose would surely be legitimate if it were approached realistically, with some recognition of our own ignorance and the scale of the task. It should start from some serious enquiry about the tools, aims and capacities involved. It would mean investigating first the legitimate and illegitimate workings of the human imagination, the way in which we organize our own purposes, and our moral relation to the biosphere we live in.

To do this, however, it would have to start modestly by examining that given human centre, by looking into our own thoughts and the affairs with which we are familiar. There would be no assurance at all of directly detecting by science the grand history of the whole. That, however, is what is at present projected. Vast and gratifying conclusions about cosmic matters are drawn directly from very slender theoretic arguments, arguments that are often scarcely scrutinized because they peep out only briefly, like very early mammals, from a protective thicket of equations.

2

PROPHECIES, MARXIST AND ANTHROPIC

THE STORY SO FAR: HUMAN SUPREMACY AND IMMORTALITY

We will discuss these odd claims in more detail later. It may, however, be best to sketch briefly at this point the story that the more recent ones tell, putting them together in a composite account. The authors involved differ in emphasis. Not all of them make all the claims gathered here. But most of them acknowledge one another with approval, and there have lately been deliberate attempts to combine them. This is now promoted as a single story, and as one that can form part of official science.

Its theme is that the human race – more properly called MAN – will colonize space far more radically and completely than has been so far expected. There will not just be a few scattered settlements in neighbouring celestial regions. Instead, in the end, as John D. Barrow and Frank J. Tipler tell us:

> At the instant the Omega Point is reached, life will have gained control of *all* matter and forces, not only in a single universe, but in all universes whose existence is logically possible; life will have spread into *all* spatial regions in all universes which could logically exist.[1] (Authors' emphases)

The term 'life' here means only *Homo sapiens*. No other earthly life-form is considered, and extraterrestrials are flatly excluded. The logically possible universes can be ignored; they arise merely from the authors' muddle-headedness.[2] But I think we should be impolite enough to remind ourselves at this point of the extent of the 'single universe' that is to be fully occupied.

17

To get some realistic perspective on this cosmic project, I was inclined to quote here a few figures from the *Cambridge Atlas of Astronomy* about the numbers, distances and temperatures involved. These figures, however, would of course not surprise the authors of passages like the one just quoted. Cosmologists constantly talk in kiloparsecs. This familiarity, however, is just the source of their trouble. The gap between such figures and human capacities is so wide that they never think of bridging it. They do not reflect on what undertaking such an enterprise would actually involve, nor ask how it would compare (say) with an offer by ants to take over the solar system. Their imaginations are not, as might be thought, over-active but inert. Absorbed in figures and used to the cosy formulae of science-fiction, they do not visualize at all what their claims really mean. (They are not, as might be thought, themselves writing science-fiction; if they were, it would be of a flat and uninteresting kind.)

The distinguished Marxist crystallographer J.D. Bernal, who was a main source of this project, first outlined it in a book published in 1929:

> Once acclimatised to space-living, it is unlikely that man will stop until he has roamed over and colonized most of the sidereal universe [i.e. the stars], or that even this will be the end. Man will not ultimately be content to be parasitic on the stars, but will invade them and organise them for his own purposes ... The stars cannot be allowed to continue in their old way, but will be turned into efficient heat-engines ... By intelligent organization, the life of the universe could probably be prolonged to many millions of millions of times what it would be without organization.[3]

In 1929, colonization as such did not have a bad name. Today it surely does, but that has not inhibited later enthusiasts for the project. Freeman Dyson (who gratefully acknowledges Bernal as a pioneer) suggests a more up-to-date form of the scheme:

> Supposing that we discover the universe to be naturally closed and doomed to collapse, is it conceivable that by intelligent intervention, converting matter into radiation and causing energy to flow purposefully on a cosmic scale, we could break open a closed universe and change the

topology of space-time so that only a part of it would collapse and another part of it would expand forever? I do not know the answer to this question.[4]

(The use of the pronoun 'we' in such proposals is intriguing.) Again, John D. Barrow and Frank J. Tipler, also enthusiastically accepting Bernal as their pioneer, update his scheme in more detail:

> If intelligent life were operating on a cosmic scale before any black holes approach their explosive state, these beings could intervene to keep the black holes from exploding by dumping matter down the black hole, at least in a short-lived, closed universe. Thus ultimately life exists in order to prevent the Universe from destroying itself! We emphasize that we do not really want to defend this possibility, but we mention it to show that it is possible that intelligent life could play an essential global role in the universe.[5]

The precaution of saying that one does not really want to defend a particular possibility has only limited effect here. The authors are still claiming that these processes are *possible*. Since nobody supposes in any case that they are actual, the full burden of justifying that claim still remains. Barrow and Tipler do indeed warn us, as we shall shortly see, that scientists sometimes put forward propositions which they themselves do not actually believe, and they seem to think it rather naive of readers to expect otherwise. But the schemes just quoted do not carry a health warning to show that they fall among these unserious suggestions, so we do not know whether we, the readers, are expected to go to the trouble of believing in them or not.

However, to resume the story – 'Life', then, meaning ourselves, or rather our mechanized successors, will not only invade all these regions, but will bring them totally under control and possibly alter the destiny of the whole profoundly. In order to start this process, we are called on to 'examine how intelligent life may be able to guide the physical development of the universe for its own purposes', how it may 'succeed in molding the universe'.[6] In case we might doubt our power to do this, Paul Davies (though he rejects much of the story) reassures us that 'we might even be able to manipulate the dimensions of space itself, creating bizarre artificial worlds with unimaginable properties. *Truly, we should be lords of the universe*[7] (Emphasis mine).

This lively prospect has, however, its price. People must transfer their consciousness from organic bodies to machines, and then to increasingly subtilized matter such as stellar dust, or perhaps light. Anyhow, as Bernal explains, 'Bodies at this time would be left far behind.' 'The new man must appear to those who have not contemplated him as a strange and monstrous creature, but he is only the logical outcome of the type of humanity that exists at present.'[8] The job may either be done mechanically or perhaps by genetic engineering. 'It is conceivable that in another 10^{10} years, life could evolve away from flesh and blood and become embodied in an interstellar black cloud (Hoyle 1957) or in a sentient computer' says Dyson.[9] This kind of separation of mind from body, long a commonplace of science-fiction, is held now to have moved from fiction to science proper, since minds can now be regarded as software which can always be shifted to other hardware.[10]

Steven Frautschi, worried about power sources for the later stages of the project, makes a helpful suggestion. 'We' might, he says,

> turn to black holes as the free energy source, and envision how life might attempt to maintain itself indefinitely, and even play a major role in shaping the universe. A sufficiently resourceful intelligence inhabiting a critical universe learns how to move black holes, bringing them together from increasingly widely separated locations and merging them to increase the entropy . . . Intelligent life might inhabit a shell of radius $R_s \sim t^{1/3}$ surrounding the black hole.[11]

Frautschi concludes regretfully that this particular scheme probably won't work, but that something like it might. In any case we ought to be working on it. 'It stands as a challenge for the future to find dematerialized modes of organization (based on dust clouds or on an e^+e^- plasma) capable of self-replication.'[12] more organization

WHITHER AWAY?

[handwritten margin notes: Do humans desire to be immortal? - medical development]

What is the point of this whole surprising exercise? For Freeman Dyson, and apparently for Frautschi, it is to achieve a kind of immortality, making it possible for these strange post-human entities who must count as our successors to stay in business – even if very feebly and slowly – after the Heat Death of the rest of the cosmos and in some sense for ever. Doubts about the concept of endless time make the for-everness problematic, and much of the discussion is about the nature of time, but Dyson hopes to beat these obstacles. Like Steven Weinberg, he thinks that the prospect of an eventual end to human life, however distant, is so awful as to deprive life now of all meaning. And the belief that some kind of post-human being, somehow produced by us, will in some sense survive seems to him enough to render it meaningful again.

Other theorists are less specific about the point of the scheme. They tend to treat the whole project simply as given, as 'the future'. In science-fiction, this word 'future' has long been familiar as meaning something highly mechanized which is both glorious and certain. The theorists we are now discussing are, however, much more boldly literal than most of their forerunners. They explicitly insist that their work is not just speculation, but part of official science. Thus Barrow and Tipler write:[13]

> The study of the survival and the behaviour of life in the far future became a branch of physics with the publication in 1979 of a paper by Freeman J. Dyson, entitled 'Time Without End; Physics and Biology in an Open Universe' [the one just quoted] ... Although the papers on life in the far future are not numerous, they have shown the progression required of physical science; the papers subsequent to Dyson's first article built on, improved and corrected their predecessors, and the discussion is now based entirely on the laws of physics and computer theory.

The standard suggested here for judging whether a topic is part of science is surely remarkable. Apparently, we need only ask whether professional scientists are publishing it in their accustomed language in a normal-looking journal and are using certain methods belonging to the physical sciences. We do not have to ask any questions about the nature of the topic itself or the kind of methods it seems to call for. Once it is in the journals, the story has become scientific; it now relates a matter of fact.

THE QUEST FOR IMPARTIALITY

To understand this suggestion, it may be worth while looking at a surprising passage in the Introduction to Barrow and Tipler's book, which illuminates what making a topic scientific is thought to involve. The authors write that they are

> cosmologists, not philosophers. This has one very important consequence which the average reader should bear in mind. Whereas many philosophers and theologians appear to possess an emotional attachment to their theories and ideas which requires them to believe them, scientists tend to regard their ideas differently. They are interested in formulating many logically consistent possibilities, leaving any judgment regarding their truth to observation. The authors are no exception to this rule, and it would be unwise of the reader to draw any wider conclusions about the authors' views from what they may read here.[14]

[handwritten margin note: scientists are more objective - open to criticism]

This innocent confidence that the speculations just outlined involve no 'emotional attachment', no bias or wish-fulfilment, seems impressive. Is it wise to be so unsuspicious on that question? But the main point is the general claim about the detachment of scientific writers. What does this mean?

Are scientists, unlike people in the Arts Faculty, never biased in their work by irrelevant considerations? To claim this would surely represent a triumph of hope over experience unlikely for people who, like these authors, have spent many years reading scientific writings. Do they then mean that scientists have no duty to take seriously the things they put in print? Can such writers always turn round and say, 'why did you bother with the arguments in my book? Of course I didn't believe a word of them'? The natural reply to that would surely be 'then why are you wasting our time?' There are indeed irritating people, both in the arts and the sciences, who often argue for positions that they do not take seriously. But, in both areas equally, they are generally recognized as a pest.

22

What is really being said here is, I think, something different, more confused but more interesting. Barrow and Tipler are struck by the fact that the subject-matter of the physical sciences is often so remote from human concerns that it does not, of itself, obviously provoke a bias. Chemists can indeed be partial, but their partiality is not usually a matter of favouring carbon over hydrogen or one protein over another. Mostly, it springs from some outside social or personal consideration.

In metaphysics, by contrast, we really may have direct preferences about such things as cosmic purpose, or how causality works, or the relation of mind to body. *What these authors hope to do is to import into metaphysics the kind of impartiality that comes naturally in physical science, simply by handling it with scientific methods.* The reasons why this can't work are of great interest, and we shall be looking at them in some detail.

For instance, it may be obvious that there is something wrong with these authors' suggestion that they can leave any judgments regarding the truth of their theories 'to observation'. Large-scale metaphysical theories such as those dealt with in this book cannot be tested by observation. They are judged by their coherence with the rest of thought, by their helpfulness in organizing it, and by their fertility.

The main point, however, is that it is not possible to break through difficulties inevitably belonging to the subject-matter of metaphysics merely by using methods lifted from a subject-matter which does not raise those difficulties. The problem of bias in metaphysics cannot be evaded; it must be met head-on. We need to start by becoming clearly aware of our metaphysical preferences, by analysing what Kant called 'metaphysics as natural disposition'.[15] We have to understand its pointings, to articulate the reasons behind them, to grasp the conflicts to which these reasons give rise, before we can analyse or face the choices they present to us.

Our own capacities and intellectual temper are not an irrelevant intrusion into metaphysics. They are a primary and well-known part of its subject-matter. The great metaphysicians have not been people trained to ignore these things. They have been ones who have had the honesty and force of mind to detect them, and to penetrate more deeply into them than the rest of us.

Unavoidably, the resulting work deals, not with an isolated

topic, but with the whole structure of our thought and experience. Unavoidably, it involves moral choices. Because of the scale of the operation, the wide implications that such thought can have, it often becomes necessary to relate one's metaphysic to the attitudes one holds on many other topics. It is not a bad idea, incidentally, to do this in science too, and great scientists have often been people who did it. But if you do not think it necessary to make it clear even whether you believe what you are writing, you do not stand a good chance of starting on this work.

FACT OR FICTION?

The claim that the authors we are considering make to literalness for their prophecies is not entirely new. People like Wells and Olaf Stapledon did sometimes think their stories actually predicted the future, and science-fiction writers have remained uncomfortably ambivalent on the matter. But the best of them have understood, as Wells and Stapledon did, that their main aim was imaginative. They were using 'the future' as a screen on which to project timeless truths for their own age. They were *prophets* primarily in the sense in which serious poets are so – spiritual guides, people with insight about the present and the universal, rather than literal predictors. For this purpose, it no more matters whether these supposedly future events will actually happen than it does for *Hamlet* and *Macbeth* whether what they show us actually happened in the past. The point of *The Time Machine* is not that the machine would work, nor that there might be Morlocks somewhere, some day. It is that there are Morlocks here now.

It was Bernal who, by contrast, instituted the claim to be completely literal. He writes, 'I believe that this scheme is more than a bare possibility, that it, or something like it, has about an even chance of occurring.'[16] (These are surely startling odds for any bookie?) But he usually speaks of it more confidently still as simply 'the future'. This makes possible the familiar kind of moral pressure by which it becomes our duty to produce something merely because it is already predestined:

> We hold the future still timidly, but perceive it for the first
> time, as a function of our own action. Having seen it, are we
> to turn away from something that offends the very nature

of our earliest desires, or is the recognition of our new powers sufficient to change those desires into the service of the future which they will have to bring about?[17]

Bernal is telling his readers to change their aims, their desires, not for the usual kinds of reason, arising out of new circumstances or new desires, but simply because their powers have changed. He tells them to do something they do not want to do, either on the grounds that they now can do it (does 'can' somehow imply 'ought'?), or because it is going to happen anyway.

Both Bernal and J.B.S. Haldane (the great geneticist who was the movement's other founding father) were deeply committed Marxists. Their confusion about a duty-to-produce-the-inevitable was therefore not surprising. The prophetic, dazzled, apocalyptic imagery of Marxism made these proposals look much less fantastic than they do now. It was Haldane who first launched the space-colonizing project in a brief piece called 'The Last Judgment'.[18] Haldane, a lively and genial character who threw off ideas like a Roman candle, does not seem to have attached any special importance to this one. Indeed he introduced it by saying (unlike Bernal) that such long-term predictions are quite unreliable. All the same, the conclusion of his piece sets the tone for Bernal's stern moral indignation against people too short-sighted to work for his chosen project. Haldane writes:

> If it is true, as the higher religions teach, that the individual can only achieve a good life by conforming to a plan greater than his own, it is our duty to realise the possible magnitude of such a plan, whether it be God's or man's . . . Either the human mind will prove that its destiny is eternity and infinity, and that the value of the individual is negligible in comparison with that destiny, or the time will come when . . . man and all his works will perish eternally.[19]

This fervour for the distant future, extending the promise that Marxism offered for a heaven after the Revolution, and this arbitrarily offered choice of only two wild alternatives, indicate the peculiar moral climate that produced these myths. (Its better-known use was, of course, to deflect moral criticism from immediate Stalinist iniquities by treating them as necessary means for future splendour.) Haldane was also responsible for launching the colourful project of genetically engineering babies to be mechanically produced outside the womb, a move which Aldous Huxley satirized in *Brave New World*.[20] Haldane himself seems, again, to have taken no particular interest later in this idea. Bernal, by contrast, took it up seriously as necessary for the earlier stages of transforming human beings in order to prolong their lives, improve their intellectual performance and fit them to colonize space.

For better or worse, Marxism is no longer a living creed today. Speculations which might well never have been made without its characteristic wildness and confidence in the remote future must now find some other means of support. Officially, Dyson, Frautschi, Barrow and Tipler depend on no religious or political argument; they present their predictions as solidly based on science.

This remarkable claim meshes with one point that does emerge about aims, which is that the future life envisaged is that of knowledge. What mechanized neo-MAN gets for his pains in outer space will indeed be knowledge; not just science, but omniscience. Barrow and Tipler end their paragraph about the Omega Point by explaining that Life 'will have stored an infinite amount of information, including all bits of information which it is logically possible to know'. An instructive footnote adds, 'A modern-day theologian might wish to say that the totality of life at the Omega Point is omnipotent, omnipresent and omniscient.' (It would not have been hard to check whether modern-day theologians do actually think that a sensible thing to say, but this was evidently not thought a matter of interest.) It emerges that, in some sense, neo-MAN has been deified.

DOES THE COSMOS NEED US?

What does this amount to? It certainly means that accumulating this knowledge is the final aim of human life. Considering the variety of other valuable human activities, this is quite a surprising claim, but it is not all. This, it appears, is also the aim of the whole universe. The Omega Point is not just the culmination of human endeavour, it is in some sense the culmination of everything. The universe aims to become complete, but cannot do so until it is completely known by people. Indeed, before people began to know it at all, it did not even really exist. It has been retrospectively constructed by human observations, indeed, perhaps only by a few observations of certain physicists. Thus John Wheeler:

> Beginning with the big bang, the universe expands and cools. After eons of dynamic development it gives rise to observership. Acts of observer-participancy – via the mechanism of the delayed choice experiment – in turn give tangible 'reality' to the universe not only now but back to the beginning.[21]

The universe is thus a 'self-excited circuit'. This idea rests on the thought, derived by a somewhat shaky route from the Copenhagen Interpretation of quantum mechanics, that quantum events are in some respects determined by the observations made of them, and do not take place without these. The world therefore had to be such that human observers would arise within it, and it was not, properly speaking, real until they had done so. Since these observers are vital for its reality, once they have arrived it is impossible that they should vanish again. They will inevitably go on to fulfil their own, and the world's, glorious destiny at the Omega Point.

ANTHROPIC

This is the Anthropic Principle, the notion that the physical universe can in some ways be explained by assuming that it must be such as to contain people. The power of this principle depends entirely on the meaning of the innocent little word 'must'. The Weak Anthropic Principle merely uses this word to mark a point of logic and consistency, as when we say that a triangle must have three sides. On this model, the fact that we are here to ask questions entails that the universe has had the kind of history that can have made our development possible.

For instance, consistency with the physical and chemical laws currently accepted means that it must have existed for quite a long time, and was therefore bound to expand to the size it now has. Barrow and Tipler use this consideration to argue that its mere size should not make us unable to believe that we are the most important thing in it, since 'the Universe needs to be as big as it is in order to evolve just a single carbon-based life-form'(p. 3).

27

This language shows again how fatally easy it is to slip over from the Weak to the Strong Anthropic Principle, which uses this same vocabulary to say something quite different and infinitely more ambitious. That principle runs 'The universe must have those properties which allow life to develop within it at some stage in its history',[22] meaning that it actually needs to have them.

What kind of necessity is this? It does not, they say, arise from purpose, from a designer's having made it to contain life. They try to do without that teleological idea because it 'does not appear to be open either to proof or disproof and is religious in nature'. Instead, they propose an interpretation that they take to be purely scientific, one which involves 'the inclusion of quantum mechanics into the Strong Anthropic Principle'. Based on the above-quoted views of John Wheeler, this interpretation claims that 'Observers are necessary to bring the universe into being.' Because these observers take part in the manufacturing or creative process, this is called the Participatory Anthropic Principle.[23]

Again, what kind of necessity is this? It sounds like causal necessity, as when we say that extreme temperatures are necessary for a tornado, or that it is necessary to keep the oven shut throughout cooking to produce a soufflé. But then how has nearly all the history of the universe gone on quite satisfactorily without observers? With the courage of desperation, participationists meet this difficulty by positing the kind of retrospective causation that Wheeler describes. For instance, if photons from a long-vanished star reach the earth and can only be said to take definite form when they are observed after their arrival, the observer is somehow causing the distant, much earlier state of those photons to have-happened in the way that his apparatus finally registers.

We need not become involved here in the desperate tangle this account raises for physics. Even if it were accepted for its own theoretical purposes, it could not possibly do the vaster job required of it here. In proportion to the totality of past world events, or of past perceived events, still more of past quantum events, these particular observations made in quantum mechanics are very few. Does it really make sense to suppose that all options in the universe have been kept open all this time in a kind of limbo, waiting for these occasional rubber stamps to arrive? Does not something or other have to have been actually happening meanwhile?

It certainly does, and we will discuss more fully later the problems raised by this promotion of human minds to the honour of retrospectively creating the cosmos. The immediate point of interest here is the ease with which Wheeler drifts over to a different conception of necessity. He accepts that world history has in some sense been going on, but says it has not been fully real. Only when observers receive it does the universe have 'tangible reality' and become a proper universe. ('Tangible' sounds helpful, he feels, because, after all, things cannot be touched till there is someone to touch them . . . ?)

An unobserved universe might, then, presumably have a history, but only a shadowy one. It would not be a real universe, any more than a four-sided triangle would be a triangle. The need for observers seems, then, to be a logical one, imposed by consistency – not, this time, consistency with the laws of physics and chemistry, but with an extremely peculiar definition of 'reality'.

This definition of reality is indeed eccentric, and has much more to do with sceptical metaphysical ideas, such as those of Hume and Berkeley, than with any kind of physical science. It is not, I think, too crude to say right away that neither the logical nor the causal account of necessity that is used to explain the Strong Anthropic Principle makes much sense. The principle's pretensions to plausibility mostly depend on a brisk oscillation between them.

Just one more element in it should, however, be noted here, namely the proof that it is Final. This is Barrow and Tipler's own contribution to Wheeler's theory; it runs as follows:

> Suppose that for some unknown reason the SAP is true and that intelligent life must come into existence at some stage in the Universe's history. But if it dies out at our stage of development, long before it has had any measurable non-quantum influence on the Universe in the large, it is hard to see why it must have come into existence in the first place. This motivates the following generalization [?*sic*] of the SAP:

> *Final Anthropic Principle (FAP): Intelligent information-processing must come into existence in the Universe, and, once it has come into existence, it will never die out.*[24]

The startling point about this is that it lands us in the very heart of teleological country. It is not merely causal, like saying that a tornado needs extreme temperatures. It is purposive, like saying that a soufflé needs a continuously closed oven. The cook's purpose is what guarantees that the oven will be kept hot long enough to perfect the soufflé. No such purpose is present in the production of tornadoes. Why is the universe suddenly being treated as a soufflé rather than a tornado? This is one of countless passages which show how traditional teleological thinking shapes the whole project of this book, though the authors constantly cover it with a façade of science.

IS IT NEEDED?

What is the real standing of the Anthropic Principle? Though it is certainly seriously intended, many physicists dismiss it pretty sharply. Heinz Pagels, for instance, speaks for many in dropping it as 'not subject to experimental falsification' and so no true scientific theory. Most physicists and astrophysicists, he says, get on perfectly well by ignoring it:

> The influence of the anthropic principle on the development of contemporary cosmological models has been sterile; it has explained nothing, and it has even had a negative influence, as evidenced by the fact that the value of certain constants, such as the ratio of photons to nuclear particles, for which anthropic reasoning was once invoked as an explanation, can now be explained by new physical laws . . . No knowledge has been gained by the adoption of anthropic reasoning. I would opt for rejecting the anthropic principle as needless clutter in the conceptual repertoire of science.[25]

He adds, moreover, very perceptively, that at first he found it impressive, but even then felt it was somehow out of place in physics:

> The anthropic principle seems less like a principle of physics and more like a biological principle resembling Darwin's principle of natural selection, here applied to the whole universe . . . As I thought more about the anthropic principle, however, it seemed less like a grand Darwinian

30

selective principle and more like a farfetched explanation for those features of the universe which physicists cannot yet explain.

In his view, it is not physics at all, but bad biology. The supposed needs of Man were (he says) being used, much as the supposed will of God was once used, as a stopgap to cover holes in scientific thinking.

3

MINIMALISM DOES NOT WORK

PUTTING SCIENCE BACK IN CONTEXT

We cannot of course attempt to discuss here the possible scientific value of the Anthropic Principle. What concerns us now about the ambitious views just summarized is their meaning for human life – including the lives of scientists outside their speciality – and their bearing on our attitudes to science.

Here the striking point is surely the enormity of the claims made. There seems no need for this kind of megalomania. Science really is a wonderful thing, and human beings really are wonderful creatures. But there are other wonderful things and wonderful creatures in the world as well. To exalt science properly is to show it in its place among them, not to send it off to an unreal, isolated pedestal among the galaxies. The true value of science is something that is only insulted by tagging it with the offer of pie in the sky. Nor can that offer restore the meaning to life, if indeed meaning is lacking. If it was wrong for religion to make capital out of offers of this kind, it is no less wrong for science to do so.

Why, then, is the value of science not explained instead in terms more suitable to it? In recent times, the strategy of excessive modesty has somewhat blocked normal, rational explanations of it. The official view of science as negative and minimalized makes it quite hard to explain its value at all, and amputating God has removed the traditional language for doing it. In this dearth of sane and reasonable praise, sudden orgies of fantasy meet a need. They compensate. Though they may not cut too much ice among close colleagues, they are readily diffused by scientific journalists; they hit the headlines and become best-sellers. They make possible an unreal, uncontrolled glorification both of science itself and of human beings as practising it. And they produce great unwillingness to set it in the context where it actually belongs – namely, that of normal life on earth.

There is now real difficulty about doing this, real confusion about what makes science worth while. Until recently, the point did not seem obscure. Scientific sages laid their own wider conceptions of human life openly on the table and showed reason to claim a particular place in it for science. Today, such publication can be quite damaging professionally. Current convention only allows their successors to make occasional intense but rather embarrassed professions – as it were, on Sundays and in their last chapters – of a faith in science which they evidently take to be justified, but which they say is quite inexplicable and irrational. Twenty years ago, Jacques Monod, in his book *Chance and Necessity*, provided an Existentialist rhetoric for doing this, and his formulae can still be heard echoing in the work of pundits who have probably never read him.

This ostentatious irrationalism is misleading and unnecessary. There are perfectly good, stateable reasons why physical science – the ordered contemplation of the material universe – is directly important to human life, why it plays a real part in our salvation. But these reasons are also reasons against cutting it off from the rest of thought. If it is to be seen as important in itself (not just for its usefulness) that importance has to be linked with the importance of other enquiries. Science is important for exactly the same reason that the study of history or of language is important – because we are beings that need in general to *understand* the world in which we live, and our culture has chosen a way of life to which that understanding is central. All human beings need some kind of mental map to show them the structure of the world. And we in the West have placed particular confidence in mapping it through methodical, detailed study.

SCIENCE REUNIFIES ITSELF

The making and using of these maps has grown hard recently. As just suggested, this is partly because, in general, methodical study has become increasingly divided and depersonalized, which is bound to make it less usable for each individual's

33

understanding of life. The maps are being made to different standards. More and more, they are required to show fine detail correctly, less and less are they designed to show the whole territory needed for actual journeys.

In the physical sciences, however, it is also because of special, temporary factors about the conception of science itself and the relation between mind and matter. From the seventeenth century on, matter was for a time conceived as dead inert stuff, so alien to mind that there could be no question of any continuity between them. Though physics has long abandoned this kind of inert matter, the idea of physical nature as 'objective' in the sense of alien and discontinuous with life has persisted much longer. It still perpetuates Descartes' notion of a radically divided universe where a continual miracle was needed to allow minds to interact with that foreign substance, matter.

This has led to the sceptical, cautious, minimalizing notion of science already mentioned. Structural ideas connecting science with the rest of life were for a time unrealistically cut back. They are now beginning to grow again because of developments within science itself – developments that are now finally reaching the rest of us through rumours of such concepts as Chaos, of an increasing distrust of the machine model, and of new ways of thinking in thermodynamics. It turns out that matter itself habitually generates order of many kinds, leading naturally through successive stages to the kinds that make life and consciousness possible. Grasping this makes possible a conception of physics as no longer detached from, but continuous with, biology and the humanities. The timeless, totally determined physical systems built by Newton and still affirmed as universal by Einstein have proved to exist only rarely, as limiting cases in conditions of equilibrium. Elsewhere there is unpredictable fluctuation. As Ilya Prigogine and Isabelle Stengers comment:

> The artificial may be deterministic and reversible. The natural contains elements of randomness and irreversibility. This leads to a new view of matter in which matter is no longer the passive substance described in the mechanistic world-view but is associated with spontaneous activity. This change is so profound that . . . we can really speak about a new dialogue of man with nature.[1]

We begin to see how, starting from chemistry, we may

34

build complex structures, complex forms, some of which may have been the precursors of life. What seems certain is that these far-from-equilibrium phenomena illustrate an essential and unexpected property of matter; physics may henceforth describe structures as adapted to outside conditions. We meet in rather simple chemical systems a kind of prebiological adaptation mechanism . . . *From this perspective life no longer appears to oppose the 'normal' laws of physics, struggling against them to avoid its normal fate – its destruction.*[2] (Emphasis mine)

They look back carefully to examine the meaning of this change:

What are the assumptions of classical science from which we believe that science has freed itself today? Generally those centering around the basic conviction that at some level *the world is simple* and is governed by time-reversible fundamental laws. Today this appears as an excessive simplification. We may compare it to reducing buildings to piles of bricks . . . Since there is no one to build nature, we must give to its very 'bricks' – that is, to its microscopic activity – a description that accounts for the building process.[3]

This conviction of fundamental simplicity was, however, long seen as essential for rationality itself, and it has had a specially strong influence on ideas about the nature of science. Being 'scientific' appeared to involve above all an unconquerable faith in this ultimate simplicity. Prigogine and Stengers quote a passage from Ernst Mach – a great pioneer of the minimalizing campaign in the nineteenth century – to the effect that the work of science is merely to simplify our thought and to make it more economical so that it is more convenient for practice. No doubt (they comment) Mach was right that scientific thought does have this simplifying function and this practical value. But could this (they ask) be its only point, its only value? If so,

How far we have come from Newton, Leibniz and the other founders of Western science, whose ambition was to provide an intelligible frame for the physical universe! Here, science leads to interesting rules of action, but no more.[4]

This strangely contracted notion has, they add, distorted and divided our imaginations by seeming to cut science off from the rest of thought, leaving us with two disconnected cultures – all the simplicity being on the scientific side and all the complexity on the humanistic one. This division is not necessary. Science itself, science proper is not isolated in this way. It is only

> classical science, considered for a certain period of time as the very symbol of cultural unity, and not science as such that led to the cultural crisis we have described. Scientists found themselves reduced to a *blind oscillation between the thunderings of 'scientific myth' and the silence of 'scientific seriousness'*, between affirming the absolute and global nature of scientific truth and retreating into a conception of scientific theory as a pragmatic recipe for effective intervention in natural processes.[5] (Emphasis mine)

THE NEED FOR MAPS

This oscillation is indeed my subject here. Both phases of it are equally misleading. For a start, the minimalism is mistaken. There is no need to copy Monod's Existentialist refusal to discuss why science is important. It is important (I am suggesting) because we need maps of how things are, and among the ways things are, the general constitution of the physical world is a prominent feature. Without being specialized scholars, we do need to grasp the general outcome of research, the shape of thought, and that shape makes a real difference to our lives. For instance, everybody's way of thinking has been affected by the Copernican Revolution that made the sun seem central instead of the earth, and also by the nineteenth-century theories of evolution.

It is not only scientific theories, however, that have this kind of importance. The same is true of the various developments in historical thinking during the same period – for instance the improved sense of the past which put the Greek, Roman and Hebrew experiences into a wider perspective, which lit up the vast dimensions of change, and which made it clear that civilization did not start yesterday. It is notably true of geography and anthropology. Even more obviously, it is also true of traditional philosophy, which has always dealt in these conceptual maps. These considerations justify, not just physical

science in particular, but 'science' in general in the older sense
– systematic understanding of every kind.

In the modern world, however, the splitting of the intellectual
scene has made map-making much harder. The great increase
in specialized learning has been paid for by a dismal habit
of competition, rather than co-operation, between its various
branches. In particular, war has far too often been declared
between 'science' in general and the arts or humanities. In the
early stages the humanities had the advantage of possession and
abused it. No doubt this is why that warfare has generated a
quite special investment from the other side in 'scientism', the
idea of *salvation through science alone*. It is this tradition that
has recently flowered into the strange fantasies which have led
me to my present subject. I don't think we need worry about
the possibility of similar chauvinistic hype for the humanities
because, if it were launched, nobody today would take the
slightest notice of it. This is certainly just as well.

The unlucky history of controversy does, however, make a
special difficulty for books like this one. The sort of objections
I am raising to distorted exaltations of science are liable to strike
some readers as attacks on science itself. Serious scientists,
however, are quite as disturbed as I am about this kind of
fame. It is actually quite dangerous for any study to get an
overblown reputation based on hype of this kind. Studies
that promise salvation, studies that are crowned as 'queen of
the sciences', as theology was in the middle ages, or treated
as sufficient for education, as classics was later, pay a heavy
price in public disillusion and resentment. We do not need to
esteem science less. What we need is to esteem it in the right
way. Especially we need to stop isolating it artificially from the
rest of our mental life.

MONOD AND THE CULT OF THE IRRATIONAL

The paradoxical, oscillating ambivalent attitude to science just
described is well expressed in the work of Jacques Monod,
because he was one of the last prophets bold and exuberant
enough to express the view freely and without inhibition.

Monod contended on the one hand that science was something much larger and more influential than a mere increase in
knowledge or understanding. It was the only possible basis for

morality in the modern age, the sole remaining source of value in a world otherwise chillingly value-free. (This is Monod 1). But he also claimed, on the other hand (Monod 2), that science was an extremely modest and limited affair, professing only to establish facts and refusing to explain them. The kind of explanation that he specially wanted to exclude from it was of course explanation by reference to purpose – teleology – which he called 'animism'. In the world he described, purpose had no place. Yet he ended his book by confidently fixing the single purpose of human life – namely, scientific knowledge. Its credentials were, he said, unanswerable because it was 'axiomatic' for science itself.[6]

There are many instructive muddles here, some of which we shall need to look at later. But it seems best to start with the difficulties about the claim of Monod 2. The minimalizing project – the shrinking of science to a modest fact-finding agency that offers no explanations – may look the easier of the two conflicting policies, but in fact it is radically impossible. Views about facts never stand alone. They are always shaped by background world-pictures which are often scarcely noticed, but which link them in a pattern and so to some extent explain them. And these world-pictures are themselves not value-free; they are always more or less dramatized.

THE CONCEPTUAL NECESSITY OF DRAMA

If examples of this insidious crypto-dramatization are wanted, the apocalyptic fantasies already mentioned might serve. But the habit is far more widespread. Thinkers like Monod himself who suppose themselves to be exposing it are as subject to it as anybody else. They are only dealing in different dramas.

The trouble is not just that they are too feeble to be properly impartial, and need more heroism to complete the job – perhaps more cold baths, or practice in the martial arts . . . ? Nor is it that we must provide for the heroism by letting machines which don't have these distressing human weaknesses do our thinking for us.

The tendency of thought to take dramatic forms is not just a regrettable chance weakness, any more than it is a chance weakness that we walk on our feet instead of our hands, or a chance weakness of our eyes that we see things in a kind of perspective which makes distant objects look smaller. What I am calling *dramatization – an arrangement of items in some sort of pattern related to their importance – is a necessary feature of explanation itself, at least in any form in which human beings can use it. And the idea of importance makes little sense unless it contains somewhere a reference to purpose.*

This means that dramatization itself is not dangerous, any more than perspective is dangerous, or indeed than breathing air or walking on two legs is dangerous. (All these things do have their dangers, but it isn't helpful to try to get rid of them altogether.) Definite views on what is and isn't important, and on the kind of life-position from which thought should start, are a precondition of all thinking. What *is* dangerous is not being aware of these views. We shall be noting that danger repeatedly in the uncontrolled dramatizations that often infest the work of those very writers who most noisily claim objectivity.

EXAMPLE: MONOD'S METAPHYSICAL ATOMISM

To show that these uncontrolled dramas are not just irrelevant consolations for exhausted thinkers, like cream buns or alcohol, but are organic parts of the thought, we can look at Monod's own world-picture, whose extravagances are visibly the result of misplaced attempts at parsimony.

Monod is crucially occupied in telling us that nature is 'objective', which means that it contains no values and no purposes. All the items in it are, he says, *contingent* – quite unconnected except by chance ('blind chance', which is a piece of illicit anthropomorphism for a start). The scene is thus that strange, arbitrary collection of unidentified flying objects which David Hume described when he said, 'All beings in the universe, considered in themselves, appear entirely loose and independent of one another.'[7] Hume, however, was merely talking about appearances. Monod claims to be doing something much more ambitious. He is boldly telling us what he takes to be the fact, the essential composition of the physical universe itself.

Now to say that this whole concern is due to chance is not clear at all, because so vast and general an idea of chance is scarcely meaningful. When we talk about 'chance' normally we are contrasting things not connected by a special kind of link – either of cause or purpose – with things which are so connected. The idea that everything, equally, is really disconnected has no

39

obvious meaning. It takes us to a level beneath all appearances, and it supplies no deeper conception of reality which might give a point to that downward journey.

Hume wanted to argue that the world of our experience consists of essentially separate atoms such as sense-data, which, 'considered in themselves', had no real connexion and might easily never have come together at all. He thought this scheme was economical. But since our experience never presents us with anything remotely like these atoms or this lack of connexion – since it always takes place in a context – this is not economical. The invention of such separate units constitutes a wild piece of constructive metaphysics.

Transferred by Monod to the outside world the notion grows yet wilder. That world, also and equally, always presents itself to us in fairly large, partially ordered chunks – trees, stars, frogs, rivers, volcanoes, people – which our thought both builds into larger pictures and breaks up into smaller pieces for analysis. The smaller pieces do not, however, according to current physics, really consist of the inert, disconnected little loose billiard-balls that seventeenth-century atomists imagined. Instead, connexion is essential to their very being. The 'particles' posited today are by their very nature elements in a system. Though they may move from one grouping to another, they are defined largely by their interactions with one another. It is not clear, therefore, what could be meant by saying that the physical world was contingent in the sense of being put together by chance.

It is interesting to notice how dependent Monod's picture of universal contingency is on the presence of those once-popular permanent, separate, impenetrable atoms. They, and indeed the whole notion of brute 'stuff' in the universe, have lately been subject to an increasing run of bad luck. As Prigogine and Stengers remark:

> Today interest is shifting from substance to relation, to communication, to time.
>
> This change of perspective is not the result of some arbitrary decision. In physics it was forced upon us by new discoveries no one could have foreseen. Who would have expected that most (and perhaps all) elementary particles would prove to be unstable?[8]

This is just one more indication that the order in the world does not consist in a single, simple, basic arrangement of indestructible balls or bricks which give the real explanation of everything. Instead, it is a wide range of much less simple, interconnected patterns. Order as we perceive it at the level of everyday experience is not an illusion. It is not a mask for a quite different order at the microscopic level, and below that for real contingency, for radical disorder among distinct bricks. It is one set among others of these real patterns – subtle, complex, interconnected arrangements. Elementary particles, as much as ponds or people, are inherently unstable, transient, incomplete entities, deeply dependent for their existence on the contexts around them. But that in no way interferes with either their reality or their meaning. Prigogine and Stengers quote Eddington's remark:

> From the point of view of philosophy of science the conception associated with entropy must, I think, be ranked as the great contribution of the nineteenth century to scientific thought. It marked a reaction from the view that everything to which science need pay attention is discovered by a microscopic dissection of objects.

They comment that indeed

> The second law of thermodynamics presented the first challenge to a concept of nature that would explain away the complex and reduce it to the simplicity of some hidden world. Today, interest is shifting from substance to relation, to communication, to time.[9]

CASINO PROBLEMS

Monod's interest in contingency, however, does not centre on this causal disconnexion between the elements of matter, but on the removal of God. To say that the world is ruled by chance means, for him, above all that it was not put together by purpose, and he takes the exclusion of purpose to be a central mark of the scientific attitude. To clarify his notion of chance, he relies chiefly on various colourful metaphors which highlight this aspect. Notably, he compares the universe to a huge casino:

> Destiny is written as and while, not before, it happens . . .
> The universe was not pregnant with life nor the biosphere

with man. Our number came up in the Monte Carlo game. Is it surprising that, like the person who has just made a million at the casino, we should feel strange and a little unreal?[10]

Notice how hard it is to depersonify one's world. Clearly, Monod thinks of a casino as embodying the total lack of order and purposive connexion which he takes as the basic condition of nature. This story is a gross insult to casinos. Real casinos are not chancy things at all but highly purposeful human artefacts, devices to produce a peculiar arrangement that is never normally found in nature – namely, a calculated disorder which can baffle prediction. How artificial this kind of 'chance' is can be seen from the amount of trouble it takes to produce it. Complex techniques and ceaseless vigilance are needed to make the roulette tables (or whatever), and the people running them, follow this peculiar pattern.[11]

The same thing emerges from the difficulties experienced by the people who publish random number tables. They must carefully examine these tables so as to remove orderly sequences which constantly get into them by accident. The fact that these sequences are accidental – that is, natural – does not mean that they can be accepted. The randomness of these numbers does not really mean 'chance' at all in the sense of accident. It means aptness for a very special purpose – namely, defeating prediction.

Why, then, should the basic constitution of the universe be – as Monod claims it is – this particular, highly artificial arrangement of disconnected items? There is no obvious reason in the nature of science why it should be so. This pattern was indeed the one used when physicists thought of matter as composed of standard little chunks which acted merely by being banged against each other. But that kind of physics was not successful. The idea of a 'scientific' approach as committed to it cannot be right.

4

THE FASCINATION OF CHANCE

PSEUDO-DARWINIAN FANTASIES

The drama of chance is, however, so attractive that followers of Monod have made great efforts to extend it right across physical science, pushing the model of Darwinian natural selection into even the most unpromising areas. Thus, Peter Atkins resists the idea that the regular behaviour of light indicates any intrinsic order in the universe, confidently explaining that this is only due to a succession of random trials and errors which must have taken place in the past:

> Light automatically discovers briefest paths by trying all paths, *and automatically eradicates all traces of its explorations*, then presents itself to us as a behaviour, which we summarize as a rule.[1]

The emphasis is mine. Atkins does not explain why light should – automatically – go to so much trouble to mislead future scientists by covering its tracks, when other cosmic entities have usually not done so. He repeats the same argument on a yet larger scale to account for the apparent orderliness of the whole universe. Before space-time proper existed – indeed before anything existed – a lot of non-existent experiments were (he declares) somehow and in some sense randomly tried:

> There is really nothing, but to comprehend the nature of this nothing, the mind needs some kind of crutch ... Imagine the entities which are about to become assembled into space-time ... as being a structureless dust ... At the creation the structureless dust of points grew into

the order we now recognise as space-time . . . Space-time emerged by chance out of its own dust . . . Think of the primordial dust as swirling, and as swarming momentarily into clusters . . . Vast numbers of such still-born universes form.[2]

These, however, all vanished again, and finally the one we now have emerged. Unlike the others, it survived, but – as he stresses – merely by chance, because it happened to be the fittest to do so.

What sense this *Just So Story* can make when there is not supposed to be anything else in existence for these pot-shot universes to fit with, or what meaning indeed the word 'chance' could have when nothing else exists, must remain uncertain. I hate to be boring but the point must be made – the idea of Darwinian natural selection only makes sense inside a relatively ordered system such as a biosphere. It presupposes an ongoing process, where there are specific conditions to meet and specific competitors who must also meet them. More widely, the idea of chance itself only makes sense within some specific existing order, an order to which it constitutes a partial or apparent exception.

More generally still, I am indeed suggesting that *all* coherent thought about the world presupposes a background of some kind of order. Whatever may be thought of that more general idea, however, the points just made about these two concepts are surely not doubtful. What Atkins has done is to mix the familiar image of cosmic dust, swirling to form nebulae, with the equally familiar one of competition between species so as to establish natural selection outside the biosphere. He then attributes this odd behaviour to objects which are not real but only potential – 'entities which are about to assemble into space-time'.

These entities are clearly yet one more distressing offspring of the current loose talk about 'possible worlds'. This phrase, when it was first used (for instance by Leibniz) had the perfectly proper sense of unrealized possibilities – imaginary worlds which might have existed, but which in fact do not. The trouble is that it sounds like the name for a special, mysterious kind of world. In the last two decades, the idea that vast clouds of such ghostly worlds somehow half-exist and do things which can matter to the existing world has become a thriving fancy. (Notices can sometimes be seen on motorways saying 'Possible Vehicles on Road Ahead', and these could doubtless be interpreted in the same way.)

What is chance really supposed to mean here? Atkins constantly treats Chaos as a positive force guiding the world in a remarkably full sense, performing many of the roles formerly attributed to God, and seems to regard it as simply a form of Chance:

> Chaos both drives and restrains the world . . . [it is] both the carrot and the cart. If everything, both structure and change, is the outcome of chance orchestrations of chaos, there must be chains linking the superficial and the deep . . . The whole course of evolution can be regarded as a geared and co-operative dissipation of energy . . . Molecules did not aim at reproduction . . .

> Since reactions are aspects of chaos, perceptions, decisions and reflections are also ultimately driven by an underlying tendency to chaos . . . All the processes in the sequence [of perception to action] are driven forward by the chaos they unleash . . . The whole of our personal history, so long as our cells survive, channels the ramifications of chaos.[3]

This extraordinary mixture of strong teleological language with inflationary misuse of the concept of Chaos marks a fairly complete bankruptcy of real explanation.

CASINO DREAMS

Why should this pattern of ultimate contingency still prove so attractive? To understand this, I think we must look at the dramas, the dreams, underlying the metaphors so freely used to explain it. The casino metaphor does not actually depict an 'objective' lifeless world containing no purposes. Instead, it shows the familiar, threatening world of fatalism. It unmistakably implies the presence of the croupier and the other gamblers, opponents who have the quite clear purpose of getting the players' money. Might the croupier also be a benign being who wants to give the punters a good time? He might, but that is not the drama Atkins and Monod want. Monod goes on:

> Man must at last wake out of his millenary [sic] dream and

45

discover his total solitude, his fundamental isolation. He must realise that, *like a gypsy*, he lives on the boundary of an alien world, a world that is deaf to his music, and as indifferent to his hopes as it is to his sufferings or his crimes.[4] (Emphasis mine)

Again, this is not a world cleared of purposes. For gypsies, as for players in a casino, the trouble is not that there are no purposes around them, but that there are hostile purposes. If the settled people among whom the gypsies camp are deaf and indifferent to the gypsies' feelings, this produces a special kind of painful human conflict, quite different from what they would feel if they were ignored by a river or a mountain. In using this image to show that our context is 'alien' Monod is effectively saying the same thing as Steven Weinberg, who laments (on similar grounds) that we live in an 'overwhelmingly hostile universe'.[5] Gypsies no doubt do sometimes feel like this. But a hostile universe is not one free from purpose.

Nature, then, stands to us in the relation of a predatory croupier or of an alien populace that resents our presence. But still another image emerges. 'We', this lonely and threatened group, are not just the human race, but the whole of life. 'The universe' says Monod, 'was not pregnant with life', and again,

> For modern theory, *evolution is not a property of living beings*, since it stems from the very imperfections of the conserving mechanism . . . The same source of fortuitous perturbations which in a non-living . . . system would gradually lead to the disintegration of all structure, is the progenitor of evolution in the biosphere . . . thanks to the replicative structure of DNA, that registry of chance, that tone-deaf conservatory where the noise is preserved along with the music. (Monod's emphasis)[6]

Life, in fact, is simply an unfortunate series of mistakes, arising from one initial mistake and expanding merely because of chronic faults in the system. It is noise, mess, a failure of the ordered scheme which was designed to prevent its emergence.

How's that for a universe without purpose? The purpose, it now turns out, was contraceptive, and the device was faulty. This failure is not recorded 'objectively' by the scientist, but quite emotively. There is real disgust at the messiness. (This tone is echoed still more strongly by Monod's followers, notably by Atkins, who writes, 'Every action is corruption, and every restoration contributes to degradation'; 'Even free-will is ultimately corruption.')[7] But the imagery has yet another point. The phrase 'the universe was not pregnant with us' cannot fail (can it?) to personify Nature once more, this time in the familiar role of a mother. But she is now an anti-mother, a being who refused to be our mother and who (it seems) became so only by mistake and quite against all her plans. She is a mother to whom we owe nothing and with whom we need acknowledge no real kinship.

Am I reading too much into Monod's language? Much more of this tradition has to be displayed before that question can be settled. Monod's imagery has some ancient roots, but it belongs of course mainly to the modern world. As far as science goes, its story starts from the attempt of seventeenth-century scientists to conceive of matter as inert, passive, standardized, unspontaneous, entirely lacking in creative powers and qualities, and to define the 'scientific' approach as one wholly committed to that conception. This project has generated a number of very strange dramas, although it was intended, from the start, to demystify and dedramatize the world-picture once and for all – to depersonify 'Nature'. In doing this, the key concept used in our own century has been the very ambiguous one of 'objectivity'. Before leaving Monod, let us have one more look at the problems this raises.

OBJECTIVITY, FRAGMENTATION AND ABSURDITY

What does objectivity mean? At a harmless, everyday level, objectivity just means keeping irrelevant biases out of science. It means being fair to one's opponents, not letting one's political views interfere with one's reasoning and again, not letting one's pride stop one giving proper attention to earwigs or tapeworms.

Even this fairly unambitious kind of objectivity is not always straightforward, because it is not always easy to decide which biases are irrelevant. But, besides that difficulty, for Monod and similar writers, the ideal of objectivity seems to have a further and quite different meaning – namely, that you should treat what you study as itself an object. It is not just that the scientists must be objective; as he constantly says, 'Nature is objective'.

This idea is by no means so clear or so harmless. When Descartes sharply divided the world into mind and matter, subjects and objects, the idea of an object took on a special negative, exclusive meaning. The two kinds of item were taken to be so different that they could have no properties in common. We ourselves were pure subjects – minds. The physical world, by contrast, was a pure object, and could not, in spite of appearances, be in any way akin to us. 'Nature' cannot share any subjective properties such as life, creativity or an inborn tendency to order.

This demand to treat nature as pure object meant that even natural beings that were alive should in effect be treated as dead, and that no part of nature, whether living or non-living, should be considered as following any kind of rational order that could flow from mind. Nothing had any mysterious intrinsic properties. Animals, as much as plants, were automata. I record this view as if it were something from the past, because it came from Descartes. But it is evidently still active. Peter Atkins says firmly, 'Inanimate things are innately simple. That is one more step along the path to the view that *animate things, being innately inanimate, are innately simple too.*'[8] (Emphasis mine. What, by the way, could 'innately' possibly mean in such a context?) This is evidently Monod's view too, though neither he nor Atkins mentions animals.

In France, where Descartes's message has been deeply absorbed, this idea still has great imaginative force. It lies at the heart of Sartre's peculiar vision of the natural world as absurdly loathsome, expressed, for instance, at the climax of his novel *Nausea*:

> I was in the municipal park just now. The root of the chestnut tree plunged into the ground underneath my bench. I no longer remembered that it was a root ... I was sitting, slightly bent, my head bowed, alone in front of that black, knotty mass, which was utterly crude and frightened me. And then I had this revelation.
>
> It took my breath away. Never, until these last few days, had I suspected what it means to 'exist' ... We ... trees, midnight-blue pillars, ... a red-haired man digesting on

a bench ... were a heap of existents inconvenienced, embarrassed by ourselves, we hadn't the slightest reason for being there, any of us ... I dreamed vaguely of killing myself, to destroy at least one of these superfluous existences. But my death itself would have been superfluous ... The essential thing is contingency. I mean that, by definition, existence is not necessity. To exist is simply to *be there*; what exists appears, lets itself be *encountered*, but you can never *deduce* it. There are people, I believe, who have understood this. Only they have tried to overcome this contingency by inventing a necessary, causal being. But no necessary being can explain existence; contingency is not an illusion, an appearance which can be dissipated; it is absolute, and consequently perfect gratuitousness.[9]

It is surely remarkable that this horrified sense of disconnectedness starts from a reaction to a *root* – something which, one would suppose, even a distracted metaphysician could see to be connected. But Sartre's despair is indeed what follows from accepting Descartes's sharp antithesis between a world of mind ruled entirely by rigid deductive necessity – guaranteed by God – and brute, total contingency in the realm of matter. Biological connectedness simply has no place here. Once you remove God, or even turn to a different kind of God who is not primarily a personification of deductive logic, this system falls to pieces entirely.

SUBJECTS IN FREE FALL

The ideal of objectivity which we are considering asks scientists, then, both to be objective themselves – in the sense of being fair – and to believe that nature is objective, in the sense of lifeless, inert and without any tendency to pattern. These are completely distinct notions. It may be that people confuse them because they think the idea of fairness justifies the belief that nature is unpatterned. Perhaps an unpatterned, wholly contingent universe may seem a fairer arrangement than one where some patterns would be built in at the expense of others ...? Theorists like Monod who stress contingency hold, with Hume, that in principle anything can be the cause of anything. This thought has clearly often presented itself as the correction of bias, a properly open-minded attitude to all possible competing possibilities. But this is not very clear thinking. We cannot keep our minds open for ever; we have to start forming an opinion somewhere.

We shall see the harm that the over-confident, unselfcritical pretension to 'objectivity' in this mixed sense has done within the physical sciences themselves – where these ways of thinking were often unsuitable – and, still more obviously, how it has distorted ideas about the relation between human beings and the rest of nature. The idea of a human subject as something standing right outside and above all natural processes is not realistic. And it gets much odder in the absence of explicit religion. When this soul or subject ceased to be seen as falling under divine discipline within the Christian hierarchy, the grandiosity became quite uncontrolled.

As for the attempt to demythologize nature, it succeeded in outlawing from science the image of a kindly, all-providing mother, a goddess who ought to be revered. But it went on, with surprising speed, to substitute for that image the much more sinister one of an enemy to be conquered – an obstruction to be trampled – or, more emphatically and particularly, of a woman to be raped. This thought, too, has given rise to a range of myths which we shall have to examine in chapters 7 and 8.

5

THE FUNCTION OF FAITH

WHY SALVATION?

So far, we have been noting the high esteem in which science has been held, and some of the reasons, good and bad, that can be given for that high esteem. We have seen the difficulty of combining high esteem with the very modest claims which scientists have recently made about the function of science, and also some difficulties that arise from trying to crush science within that modest function. We have seen, for instance, how hard Jacques Monod found it to take his own medicine, to depict a depersonified, non-dramatized, purely factual, 'objective' universe. And we have glanced at some startling, supposedly scientific predictions which are now thriving as compensations for the unreal modesty of this official view.

You may still, however, be wondering why, in my title, I connect that high esteem with the rather colourful word 'salvation'? Why not just talk about the *value* of science? As I have suggested, the point of using stronger, less everyday language is to show how much the whole thing matters, and especially to draw attention to the high ambitions underlying strong claims about that value. These are the claims that have brought science into a competitive relation with religion.

It is, or course, quite widely believed that science and religion inevitably are in conflict. Many people, indeed, suppose that this battle has already been won – that science has in some sense 'disproved' religion, and reigns instead of it. This is an extremely odd idea, since it has to mean that they have somewhere been competing for the same job, and it is not obvious what that job might be.[1] If science were really no more

51

than the modest, narrow, detailed, negative memory-bank that Monod 2 described, it could not possibly come into conflict with any other branch of thought at all, let alone disprove anything outside its own borders. And it certainly could have no relevance to religion.

The idea of a conflict arose at the end of the last century, when scientists still meant by science something much wider than a memory-bank. They included in it a whole myth, a philosophical conception of the world and the forces within it, directly related to the meaning of human life. They saw this penumbra as part of science because it was needed if scientific propositions were to have their full bearing on the rest of thought. Unlike their forebears, however, these late Victorian scientists were beginning to see it also as an entity on its own, something cut off from, and perhaps hostile to, other ways of thinking. People like T.H. Huxley meant by science a vast interpretative scheme which could shape the spiritual life, a faith by which people might live. This faith was a competitor with existing religious faiths, not a way of having no faith at all. The reason why science and religion seemed to come into competition was that they were seen as rival attempts to do the same work.

The question is, which work?

RELIGION INVADING SCIENCE

The less interesting source of clashes is the one which has had the most attention. It is where (crudely speaking) religion tries to do the work of science by claiming to establish physical facts. But this still only produces competition if those facts are ones viewed as specially important – if symbolism makes them look crucial for human destiny. Thus, questions about the earth's flatness or roundness have never raised the same passions as those about its central position and its motion or stillness. (Creationists do not usually seem much concerned to restore the flat earth, with a hard, domed 'firmament' or ceiling above it and waters beyond both, which is plainly described in Genesis.)

The trouble, then, is not just that religious authorities are attempting to intrude on science. It is that scientific findings are already impinging on life, so that some way must be found of digesting their meaning. This happens most readily over large scientific changes affecting our conception of what it is to live and – more especially – to die. Robert Jay Lifton, examining our range of responses to the challenge of human death, suggests that in recent times the issue of death has been central. The alarming point about the theory of evolution was, he thinks, not so much its undignified account of the human past, but its transference of future hopes away from personal immortality to continued earthly development, and above all to the development of science itself. As he remarks,

> The biological mode [of conceiving immortality] in no way began with Darwinism – it is perhaps the most fundamental of all images of immortality . . . But, as the imagery took hold, man's sense of biological continuity was extended back into the infinite past and therefore . . . into the infinite future . . . It was a coming-of-age of the scientific endeavour to assert its version of the mode of immortality via man's works.

There was, he thinks, no necessity to see this sense of earthly continuity and the religious one as competing alternatives:

> Of course many have retained both scientific and religious modes of immortality in their personal symbolic life, including Darwin himself. The either/or image was made possible only by a literalized version of the religious mode, a view of Genesis and the Bible in general as, in every respect, literal recordings of actual events . . . *The real non sequitur in any context is precisely such literalization and accompanying either/or assumptions.* Nor is the scientific endeavour in any way free from the danger of an equivalent literalization, the holding to particular ideas, principles or images beyond their relevant context . . . What we call dogma, in this sense, is an attempt to 'stop time', to 'stop history', to stop the flow – the perpetual reconstruction – of collective expressions of truth, meaning and human connexion.[2] (Emphasis mine)

What produces the impression of competition is (he suggests) the insistence on clinging to discredited factual beliefs as literally true – an insistence that flows from the failure to shift the symbolism, failure to rethink the underlying spiritual truth in the new terms that are needed. But this fault (as he points out) is not confined to the religious side of the debate. That same literalization – that attempt to meet spiritual questions by a set of allegedly factual propositions – occurs also among those who champion science. It is exemplified by the quasi-scientific future-fantasies we have already glanced at.

Efforts to meet the religious needs in scientific terms by providing new symbols are understandable because those spiritual needs are real and urgent. It is creditable in scientists to notice that they do have to be met. But, as Lifton stresses, there is no short way of doing this. Such needs are met through a slow and painful communal development, through the effort to find, in experience, new effective symbols, which must grow out of better ways of living and feeling.

This emergency is, incurably, a moral and imaginative one. Hardware will not help it. It does have an intellectual aspect, but one that involves understanding the human imagination rather than predicting the progress of the cosmic Heat Death. What is needed when new scientific facts clash with beliefs formerly held significant is not to declare war, nor to bend the facts. It is to rethink the significance, to look much deeper into what underlies the symbols.

FINDING NEW VISIONS

That mode of reconciliation was open to Christians both in Galileo's and in Darwin's time, and on both occasions many of them used it. Symbolic and allegorical interpretation of apparently factual stories had always been an approved Christian tradition, freely used by the fathers. Origen, for instance, already pointed out that the seven 'days' of creation could not be taken literally, since they occurred before the sun and the moon existed.

This symbolizing approach had obvious advantages in dealing with matters admitted to be beyond human understanding. But fanciful use of allegory brought the method into discredit during the Renaissance. In Victorian times, many thinkers, using symbols more critically, managed to avoid crude, competitive debate until about the end of the nineteenth century.[3] At that time, however, those on both sides who preferred to dispute, and to link the issue to political quarrels, began to gain the upper hand. They treated the idea of literal, factual truth as the only one available, and set it up as a prize to be fought for. This was the time at which 'being scientific' began to seem synonymous with atheism. Battles are always popular, and many people, not unnaturally, find them easier to handle than the hard thinking that is needed here. This response therefore caught the public attention. But it has never been appropriate.

RE-READING THE SYMBOLS

Apart from simple emotional attachment to former symbols, the business of deliteralizing religious doctrines is hard because it involves spelling out in some detail just what the symbolic meaning has been – just what positive truth should still be saved when traditional stories stop being taken literally. This is difficult because of the richness and ambiguity of symbols. It can take a great deal of complex metaphysical thinking to explain a symbol which did not seem to need much explaining when it was simply accepted as a fact.

For example, the Copernican Revolution disturbed the general symbolism of 'up' and 'down' in a way that has proved a lasting nuisance. In theory, it might seem easy just to accept that we are still not going to fall off the earth, without worrying about what things might feel like in parts of space where we need never go. Indeed, further worry about this kind of thing could be seen as neurotic and superstitious. Yet vertigo still persists. The world's new position is often felt to have astonishing moral and metaphysical consequences, such as that truth or justice has become a meaningless concept – sometimes, even more strangely, the psychological consequence that all human kindness is an illusion.

Again, the symbolism of finite and infinite works powerfully on the imagination. Pascal was disturbed by the silence of those infinite spaces, even though, since his religious faith remained, the symbolism of the 'music of the spheres', which had been lost, might not seem too hard to salvage.

These clashes over literal beliefs about facts, then, are often more awkward than they may look. They can pose difficulties for our imagination, though there surely does seem, in principle, to be a clear way to deal with them. The factual findings of the sciences must (we think) be accepted, but the symbolic meanings which were attached to the earlier view of the facts are no business of science, and need not be affected at all by such changes. These symbolic consequences are by no means confined to religion; many of them directly concern even the most sceptical and atheistical of us. They are spiritual, moral, metaphysical or psychological in a sense of 'psychology' which falls outside today's narrow notion of science. These topics do not cease to be thought about because they lie outside the borders of science. They have to be thought about in other ways.

In general this recipe is surely right, but imaginative changes of this kind are harder to make than we might have hoped. Psychological symbols cannot be altered in the brisk way in which one might change a road-sign. They are not, like words, conventional signs, loose pieces arbitrarily nailed to their meanings. Nor are they even fixed items, standing in regularly for a single meaning, as Freud seems to have thought. For him, *pen* simply meant *penis* and *bag* meant *womb*. Questions scarcely ever arose about what the penis or womb themselves meant. In our imaginations, however, these questions are extremely important. Such symbols are not simple counters, they are gateways to whole uncharted territories.

Changes in the literal meaning of the symbols, such as the Copernican reinterpretation of up and down, force us to rethink all that they formerly symbolized. We have then to disengage literal from symbolic changes. We have somehow to stop irrelevant colouring, introduced from the new factual belief, from spreading over a whole field of meaning which we did not before have to define. In the Copernican case this has not been done at all effectively, because the natural link between the ideas of higher and better is extremely strong.

Similar, but I think much less excusable, confusion affects the dreams of human immortality at which we glanced in chapter 2. Its inventors, believing that physics and computer science now make endless life literally possible, seem to suppose that this can give us just what has been traditionally symbolized by the idea of eternal life. This is surely like interpreting the foundation of the State of Israel to mean that everything conveyed by the old phrase, 'Next year in Jerusalem . . .' can now be attained by ringing El Al for a package holiday. It is the sort of mistake that unlucky people make in the black fairy-tales, where you are granted literal fulfilment of your wishes, and realize to your horror that it is disastrously different from what you really wanted.

Is mere perpetual survival really the point at all? Does not something depend on what sort of life we expect these remote beings to have? If perhaps it were a life that we would think worse than no life at all – if it were one that they themselves would think so – what would follow? Would our foreknowledge that this fearful life would go on for ever – that perhaps, like Tithonus, they cannot even die – be the sort of thing that would supply meaning to our life now? And is the prospect of an end the sort of thing that can destroy that meaning?

SCIENCE INVADING RELIGION

We will be returning to these questions about symbolism repeatedly. They arise out of considering the other kind of possible competition between science and religion, which to my mind is much more interesting and much harder to deal with. Science can clash with religion if it – science – is in the business of providing the faith by which people live. Is it actually in that business?

This kind of faith is not primarily a belief in particular facts. It is not what William James's schoolboy meant when he remarked, 'Faith is when you believe something that you know ain't true.'[4] The faith we live by is something that you must have before you can ask whether anything is true or not. It is basic trust. It is the acceptance of a map, a perspective, a set of standards and assumptions, an enclosing vision within which facts are placed. It is a way of organizing the vast jumble of data. In our age, when that jumble is getting more and more confusing, the need for such principles of organization is not going away. It is increasing.

FAITHS AND PRINCIPLES OF SELECTION

Faiths of this kind do not need a god. This is plain from the case of Marxism, and also of Taoism. The same examples show, too, that one's faith does sometimes affect one's view about facts. It can determine which facts one is prepared to accept. Marxists have habitually accepted detailed predictions about the future, and also particular historical beliefs about the past, because these fitted well with their guiding theory. At a glance, this might seem to give us a clear distinction between this kind of faith and a belief in science. Surely, we feel, scientists accept all facts impartially, on universal standards of evidence? Surely they are never influenced in accepting or rejecting them by the demands of a particular theory?

In fact, of course, they are and must be so influenced. For instance, academic scientists today mostly refuse on principle to consider or publish any research about the topics now viewed as spooky, and grouped under the heading of parapsychology. In this case the reason is a frankly metaphysical objection to causation that cannot be explained by the laws of physics. This may be a good reason, but the point is that it operates *before* considering the evidence. It simply forbids all interest in a range of topics which equally intelligent scientists at the end of the nineteenth century found extremely interesting – because their metaphysic was different. At that time, too, scientists were willing to discuss the metaphysical issue itself, but today they have put it outside their frontiers. Similar things happen constantly even where there is no suspicion of metaphysics. For instance, the theory of continental drift was long dismissed as unscientific, and so for a time were James Lovelock's suggestions about damage to the ozone layer.

SELECTION PROBLEMS

This dismissiveness is not just a fault. There does have to be some principle of selection. There is no such thing as confronting all possibilities impartially, with no prejudices about what is initially probable. Scientists, like other people, must operate with a mental map or picture of the world which emphasizes certain areas and brings out certain lines as promising. Stories that fall far away from these lines won't be perceived as possibilities at all.

The scientists who rejected the idea of continental drift were not just being foolish. They were using a map which had no room for the possibility of unfixed continents, just as the mediaeval map did not allow the possibility of an unfixed earth. Similarly, when people like Lovelock began to suggest that human activities might be causing damage to very high atmospheric levels and even in the stratosphere, this did not seem like a possibility. People didn't believe in that sort of thing, because the principles that they did believe in made it look impossible. Lovelock comments:

> It is a scandal that the vast sums spent on expensive big science of satellite, balloon and aircraft measurement failed to predict or find the ozone hole. Worse than this, so sure were the computer programmers that they knew all that mattered about the stratosphere, they programmed the instruments aboard the satellite, that observed atmospheric ozone from above, to reject data that was substantially different from the model predictions. The instruments saw the hole, but those in charge of the experiment ignored it, saying in effect, 'Don't bother us with facts; our model knows best'. The Ozone War is littered with stories of this kind of military incompetence.[5]

It is perhaps no accident that all these cases involve imagining that something which seems fixed and permanent is really moving and changing. But of course this isn't the only kind of possibility that gets hidden. The whole point is that what is hidden might in principle be anything.

Our next thought is: these people were too narrow, too unimaginative, too wedded to their own map. This is right, but the tendency is hard to correct. One can't get good service from such a map or picture without taking it seriously, which is why I think it is relevant to speak of a *faith*. We have to be to some extent committed to our world-pictures. It is worth remembering the remarkable faith with which Darwin stuck to his new conception of biology through the many years when very little unambiguous evidence for it emerged, while the fossil record remained obstinately unwilling to support him.

At the extreme of commitment, people are quite unaware that they are using such a map or picture at all. They feel as if they are simply looking directly at the world. This condition of unselfconsciousness about one's concepts is very common, even among intelligent and well-informed people. It is where we all start, and we are only forced to become more critical – more aware of alternative schemes – by receiving a series of knocks from errors and conflicts which make us aware of other possibilities.

At the other extreme, it might in theory be possible to get so sophisticated, so over-critical that one would stop taking one's own map seriously at all. People professing relativist doctrines do sometimes talk as though they have got into this state, but actually to arrive there would presumably be terminal for their thinking. It is not possible to go on thinking if you qualify all your thoughts and words by the comment, 'or so they are apparently saying in my culture at present'. In fact, relativism is itself usually extremely selective, and amounts only to using a map on which other people's views, but not one's own, appear as relative to their culture.

Anyone who is not a Marxist can see this happening over Marxism. Anyone who is not a Mormon can see it happening over Mormonism. Very ambitious conceptual schemes like these are useful because they make this universal condition visible. With less extreme and less narrow views the bias is less obvious, but of course it is always there. We can no more do without conceptual schemes than we can do without some particular form of eyesight or some particular standpoint from which we see the world. We need it for thinking just as we need some particular language if we are to talk.

Yet it notoriously is true that, from any point, only certain things can be seen; in any language, some things cannot easily be said. The Chinese language has, it seems, no word for God, a situation which has made great difficulty for missionaries. Similarly, the sociobiological language of 'selfishness', 'spite' and 'investment', which is now constantly used in ethology – and sometimes elsewhere – for discussing motivation makes cramped and biased views on that topic unavoidable. Our only way of correcting these biases is to keep checking with people who stand somewhere different or speak different languages, and noticing the discrepancies.[6]

This work is endless. The hope of avoiding it is, I think, the great thing that has made people put faith in 'science' as the final cure for all intellectual ills. They have seen it as a way of doing without conceptual schemes altogether – an instrument for showing the world directly and impartially, without the drawbacks of any single perspective. In the seventeenth century this hope led to the search for a universal, scientific language, enabling scientists to avoid all the particular prejudices that go with the various natural languages.

This was part of the more general hope that science could itself become a kind of universal language – a single, vast ordered pattern, mirroring the pattern of the universe and revealing all truths. The still-surviving idea that there is a single 'scientific method', the same throughout all the sciences, seems to be a relic of this dream. This hope of unification is perhaps a central thought when science is hailed as the core of learning, the cure for all intellectual ills, and when the 'scientific attitude' is held to be the one that insists on using this one Procrustes' Bed for all subject-matter.

The attempt to dedramatize nature was meant to form part of this great scheme. It was intended to bypass weighted language and biased points of view, so as to reveal the facts directly. But there are, unfortunately, far more ways in which points of view can be biased than one might hope, and also many good reasons for using various kinds of language. As a distant, guiding ideal, the removal of bias is thoroughly proper. But the notion that anybody could actually achieve it has turned out no more realistic than the attempt to see the world fairly by standing nowhere.

The history of thought is littered with supposedly universal and final schemes which have had something good in them, but have failed lamentably in what they claimed. It has become clear that we can indeed aim to correct partiality by balancing one bias against another, but can never assume that we have finally succeeded in becoming universal. Our knowledge does not consist of pure bits of information, warranted final, but of world-pictures which balance each other and constantly need modification. That is why the disinfecting project has fallen into the rather dangerous difficulties that I have been describing.

6

ENLIGHTENMENT AND INFORMATION

SALVATION?

The point of using the dramatic word, 'salvation' is, then, to bring out the vital importance to all of us of these various faiths on which we all – and not just the professedly religious – depend, faiths without which we would be lost. As I have explained, I use this strong word because, if one merely talks about seeing great value in something like science, this may suggest a detached, market situation (shopping for values; perhaps waiting till the revaluation sale comes on . . . ?). At best it may suggest an art-gallery, a value-museum where we stroll through at leisure and admire the exhibits detachedly, preparing to write a critical essay about them. The point about salvation-talk is that it admits the gravity of the need. It knows we are in a pretty bad way for a start.

When G.E. Moore said that the highest human good was the 'admiring contemplation' of beautiful objects or beautiful people,[1] I think he did suggest the detached museum-attitude. I think we should resist that suggestion. Of course there is nothing wrong with museums and art-galleries. Of course it is quite possible to look at pictures or exhibits in a spirit which is an active quest for salvation rather than just a visitor's idle curiosity. Perhaps that is what Moore meant. But I don't think the words 'admiring contemplation' convey it, and neither, in general, does the current language of 'values'.

The need for salvation is something urgent and drastic. At the beginning of the *Pilgrim's Progress*, when Christian cries out 'what shall I do to be saved?' he is desperate, and he has to find an answer. What he needs is a faith, something to believe in. And when we look at the extraordinary variety of faiths that people have held, it is surely clear that this need is deep and virtually universal. What people seek, and what they will accept, naturally varies with the particular spiritual disaster that they fear. But evidently, they often do feel the threat of some serious spiritual disaster.

Intellectuals sometimes dismiss this fear as if it was childish, a mere crying out for consolation from an absent parent. This ignores the extent to which they themselves draw strength from reliance on the systems of thought that they use. The habit of doing mathematics, for instance, can certainly be used as a consoling addiction, but to dismiss it as a mere addiction would be short-sighted. Similarly, religions are often used childishly, and they do sometimes use the imagery of parents, sometimes also rather crude imagery of safety in an after-life. But that is never all that is meant. If people like President Reagan expect to be literally carried up in chariots to enjoy pie in the sky, they have missed the point.

The cry for salvation doesn't arise only in times of outward danger. In fact those times may distract people from it. It is a response above all to *confusion*. What the pilgrims are seeking seems to be above all a meaning for life, a set of connections, however incomplete, that will make some sense of it, a general shape which will bring conflicts and clashes into some perspective. That is why I think the language of separate 'values' is often unsuitable for it. This need is not one that can be met by picking up a couple of nice values here, a couple more different ones there. It concerns wider connexions, it demands some sort of wholeness. And pursuing an intellectual enquiry can itself be a move towards finding this, just as plainly as accepting a religion can.

LIGHT AND ENLIGHTENMENT

It is extremely interesting that the metaphor of light and darkness plays so big a part in both these concerns. We use the word 'enlightenment' today with equal naturalness in two very different contexts. It is the name of that predominantly anti-religious intellectual movement which gave rise to modern technology and modern learning. But it is also the name given in Buddhist thought to the state of release from the snares of the world. What, if anything, do these two senses have in common?

The link seems to be an emphasis on vision, on understanding. The Buddhist sage sees the world as it is, and understands finally what matters and what does not. The ideal modern learned person also in some sense sees the world as it is. But what does that sense involve? What are we academics actually after in our enquiries? And why do we think them so important?

The accounts of intellectual work that are most fashionable today usually answer that our aim is simply to acquire more information, facts delivered piecemeal in individual 'bits'. But it is not obvious how merely collecting information can be seen as such a central aim. Information as such is not necessarily valuable at all. The number of facts is endless, and, on anybody's view, many of them are trivial. Hawking's 'complete description of the world', if gathered, would be chiefly an endless telephone directory, unusable by any conceivable kind of enquirer. People don't normally want merely information as such; they want interesting information. They want answers to questions that are worth asking. If they are given any other kind of information, they usually can't even remember it.

This is, of course, not to say that we only want useful knowledge. Curiosity does often flow around spontaneously, raising all kinds of unaccountable questions. But when this natural curiosity is followed, it surely tends to concentrate itself on particular topics, and then to look for connexions within them, and between those areas and the rest of life. It is selective in a way that eventually does lead to the quest for system. The lifelong beetle-gatherer does not merely collect beetles, but usually also thinks about them. And even before that happens, the fact that it is selective at all – that beetle-gatherers will so sharply refuse to collect stamps instead – means that curiosity is quite unlike the indiscriminate demand for information-as-such that current theory suggests.

The mere fact that we get bored so easily by being showered with information that we don't ask for shows this clearly enough. Education would be a lot easier if this wasn't so. Curiosity considered as a mere appetite for particular facts never becomes a very strong or important motive unless it is backed by some deeper and more general wish. I am suggesting that that deeper

64

wish, making curiosity into a serious motive, is centrally the fear of lostness, the desire for connexion and meaning.

FINDING OUR BEARINGS

Human beings, then, like other intelligent creatures, direct their enquiries to things that strike them as important. They don't ask all the questions there are; they ask questions that matter to them. And what matters is surely – for everybody else as well as for Buddhists – what brings things together, what shows a pattern, what tends to make sense of the whole.

The perdition from which we want to be saved does not consist essentially in a shortage of information in our memory-banks, but in being 'lost'. From the point of view of both kinds of Enlightenment, what is disastrous is not just ignorance or even error as such, but ignorance and error about the whole and where we stand in it – failure to understand the world sufficiently to grasp our own position in relation to what matters in it.

In this tradition, to be 'damned' is to be adrift; not to know where we are. That is why the metaphor of light is so important. Light stands for an explanation of life, and an explanation satisfying, not just to our emotions nor to an impartial curiosity, but to our sense of rational order. This is bound to be particularly true in our own culture, because our Enlightenment tradition lays a particular stress on the kind of order that is satisfying to reason. Its special demand is that each person's reason should get that satisfaction directly, rather than just accepting the authority of rulers who are supposed to think for other people. As Kant put it in his sharp little 'Essay on Enlightenment':

Enlightenment is man's release from his self-incurred tutelage. Tutelage is man's inability to make use of his understanding without direction from another ... It is so easy for me not to be of age. If I have a book that understands for me, a pastor who has a conscience for me, a physician who decides my diet, and so forth, I need not trouble myself. I need not think, if I will only pay – Others will readily undertake the irksome work for me.[2]

65

OMEGA FOR INFORMATION

That emphasis on enlightenment of the understanding and the conscience is a quite different ideal from just maximizing the stores in the memory-banks. Contrast here Barrow and Tipler's account – which we glanced at briefly in chapter 2 – of what they take to be the culminating purpose, not just of human life, but of the whole cosmic process. The universe will (they say) eventually find its fulfilment in the Omega Point, at which

> life will have spread into all spatial regions in all universes which could logically exist, and will have stored an infinite amount of information, including all bits of knowledge which it is logically possible to know. And this is the end.[3]

A footnote adds 'A modern-day theologian might wish to say that the totality of life at the Omega Point is omnipotent, omniscient and omnipresent.'

Why should this be the final cosmic achievement? Life, it seems, ranks as God simply by *possessing* information – by storing it, not by doing anything with it. Why this mere possessing should have value, any more than possessing jam in a cupboard or gold in a vault, is not explained. Storing, after all, can only be a means. Things are stored to be used.

THE RESURGENCE OF COSMIC PURPOSE

Apart from this odd choice of a goal for the universe, however, why is the universe now taken to have such a goal at all? As we noticed before, the renunciation of all reference to a cosmic purpose used to be fundamental to the chief heroes and champions of modern science from Galileo to Jacques Monod. Avoiding all reference to such a purpose was taken to be a necessary sign of being 'scientific'.

This view has never been publicly and officially repealed. What has changed in the last twenty years is, as we have noted, just the practice, and that mostly in a rather furtive, unofficial sort of way – in popular books and last chapters, as in Stephen Hawking's promise that cosmological theory will help us answer the question 'why we are here' and will put us in the position to gain 'the ultimate triumph of human reason' by understanding the mind of God. Barrow and Tipler are exceptional in making a welcome stand against this casual drift. They do explicitly discuss teleology, and they claim to give good reasons for reinstating a form of it as part of science. Their reasoning about it is not impressive. But it is a lot better than no reasoning at all.

Since history invaded physics – since the image of the Big Bang took over and displaced Galileo's carefully timeless, reversible world – this kind of talk has gradually become quite common; clearly it causes no embarrassment. If we were just looking for evidence of the power of teleological or animist thinking to resurface and flourish in the most hostile and forbidding environments, this change would surely be a striking enough example. But we ought, I think, to do more than that.

We should not just notice it as evidence of an irresistible, senseless urge. We need to take it seriously and see what it means. If teleology has again become a legitimate way of thinking – legitimate enough to be used, even in last chapters – then it can't be confined arbitrarily to those uncriticized contexts. Rules have to be worked out again for its proper use. We need to develop further the distinctions which Aristotle began to sketch, between good and bad teleology, between different uses of it, between right and wrong contexts in which to use it. If we stop vetoing it altogether, what we need next is to understand its function.

KNOWLEDGE AS PART OF FREEDOM

The first thing that we need to look at in each case of teleological thinking is the world-picture, the vision being expressed. In the case just mentioned of the Anthropic Principle, the central point is surely a value-judgment about the supremacy of knowledge. Information-gathering is declared to be the most splendid of human achievements. The passage projects on to a cosmic scale a particular estimate of the relative value of various human occupations, an estimate that is crude and naive because it simply plumps for one candidate without considering any alternatives. Unless it is properly explained, it is in danger of turning out to be essentially the same value-judgment that was more modestly expressed in 1870 by a certain Dr J.H. Bennet; he wrote:

> The principal feature which appears to me to characterize the Caucasian race, to raise it immeasurably above all other races, is the power that many of its male members have of advancing the horizons of science, of penetrating beyond the existing limits of knowledge – in a word, the power of scientific discovery. I am not aware that the female members of our race participate in this mental power, in this supreme development of the human mind.[4]

Today, this attitude mostly finds more cautious expression in the apparently modest terms of treating information-gain as central to every kind of process. It is possible to do this, because the language of information is a flexible one and can be used to describe a lot of different phenomena – sometimes usefully, sometimes not. This is actually true of a host of other conceptual languages as well. But if you don't notice that, the possibility of treating so many things as information-transactions looks impressive. It begins to seem that – metaphysically – the whole world really *is* just a mass of information. Nobody is being informed and there is nothing to be informed about. Information itself is the basic substance of everything.

In that context, Barrow and Tipler's statement is not just an aberration. It is a natural outcome (if rather an exuberant one) of the idea that this collecting, this possessing of information is the whole aim of thought. I do not think that those who use this language have noticed that they are saying something revolutionary, but they are. They perceive themselves indeed as saying something perceptive, something smart and new, but still something obviously right, something reductive. Their impression is that they are just giving a clearer, more intelligible development of the Enlightenment tradition which they have grown up with, and which they probably don't want to question. But in fact they have lost touch with that tradition entirely. Merely storing information is not an Enlightenment ideal at all.

The Enlightenment value-system centred on a strong moral campaign, designed to exalt certain values – freedom, independence, activity, autonomy, honesty, moral courage – over other, more hierarchical and corporate ones such as loyalty, love, modesty, reverence and discretion. It wanted to make sure that the effective unit of morals was the individual, not the society. When it emphasized intellectual work and the quest for knowledge, it did not treat knowledge simply as an end in itself, but as a field for that enterprise. Enquiry mattered, not primarily as a source of supply for the information-store, but as a way of teaching people to think for themselves.

That is the conception that still informs our ideas about education. When we think of education as a light which dispels our darkness, what we surely have in mind is not so much that it saves us from a famine of facts as that it cures an inability to think freely about them.

This is clear too in current disputes about the role of artificial intelligence. The objection that people feel to reductive accounts of how human thought can be replaced by computer programs isn't just a lack of confidence in the storage capacity of those programs. It is a well-justified alarm about the state into which people fall when they no longer even try to handle their affairs by their own judgment. To quote from another famous Enlightenment manifesto, Mill's essay *On Liberty*:

> Supposing it were possible to get houses built, corn grown, battles fought, causes tried, and even churches erected and prayers said, *by machinery* – by automatons in human form – it would be a considerable loss to exchange for these automatons even the men and women who at present inhabit the more civilized parts of the world, and who assuredly are but starved specimens of what nature can and will produce.[5]

People need to do these things for themselves; the activity is itself the end. Mill doesn't make any exception to this for the collecting or possessing of information. In fact, this is one of his prime examples of something that is no good if it is done automatically. Thus, in defending freedom of speech, he answers the objector who says that some things need not be discussed because the truth about them is known already, by pointing out that it is not enough merely to possess the truth:

> Assuming that the true opinion abides in the mind, but abides as a prejudice, a belief independent of and proof against argument – *this is not the way in which truth ought*

to be held by a rational being. This is not knowing the truth. *Truth, thus held, is but one superstition the more . . .* Even in natural philosophy [he means physical science] there is always some other explanation possible of the same facts . . . it has to be shown why that other theory cannot be the true one. Until this is shown, and until we know how it is shown, *we do not understand the grounds of our opinion.*[6] (Emphases mine)

THE STATUS OF MORAL JUDGMENT

Obviously, I am not here to lay down these views about the value of free thought dogmatically. (That is one thing you can't do . . .) Mill and Kant may be wrong in their ideas about the importance of thinking for oneself. These are value-judgments. They can be questioned on moral grounds and of course they often have been. What I am pointing out is that these ideas about the importance of free thought – which are widely accepted in our culture and commonly used to defend academic enquiry – clash sharply with the current notion that what is good about science is just that it accumulates information.

That notion is not itself a fact. It too is a value-judgment, and one whose grounds are none too clear. If it is to be held, it needs to be explained and defended. To say that the mere acquiring of information as such has supremely high value is a startling and paradoxical claim which needs explaining. To dramatize this judgment further by claiming (even in one's last chapter) that information-gathering is the aim of the whole cosmic process is not to evade the need for justification. It is to make it still more urgent.

This is an example of what I mean by a faith. And the point about such faiths is that they are not just a harmless indulgence, an elegant amusement set aside from the rest of thought. They are a basic, active element in it which constantly affects our ideas and needs constant attention. If we don't have one kind of faith, we are very likely to have another. Faiths which are not watched grow like mushrooms in the dark. It is important, and quite difficult, to think them through and to make sure that they are of the kind we want to harbour.

Is this cult of information perhaps itself a new faith, not yet fully expressed, one that is independent of current Enlightenment-type thinking? That would be something very interesting. Of course we can part company with the Enlightenment tradition if we want to, and in many ways we have done so. There is already quite a healthy industry of Enlightenment-bashing going on. There are many current issues where we need to make new departures. This is not always because the Enlightenment sages were wrong, but because we have already gone as far as we can usefully go in the directions they suggested, and need to find new ones.

KNOWLEDGE AS UNION WITH THE DIVINE

But can this be what those who put such a high value on information are trying to do? The trouble is that *there is no other obvious modern way of thinking at all, apart from the Enlightenment one, that it likely to put anything like so high a value on knowledge.* The only other likely resource is an older tradition, the Platonic and Aristotelian one, which exalted contemplation over action as the supreme human activity. This is indeed a powerful and sophisticated line of thought. It is the one which originally launched the whole vast ship of European learning, with its countless Academies and Lyceums called after Plato's and Aristotle's schools. Its ideas have constantly worked to keep that ship afloat and to drive it forward.

But that tradition is essentially a religious one. It exalts contemplation as reverent wonder, as a means of union with the Divine. Even Aristotle, who eventually dropped all belief in a transcendent God or an immortal soul, thought that the point of knowledge was contact with the rational order of the universe, an immanent, divine order which was something distinct from ourselves and above us, the Unmoved Mover which is the ultimate object of our love as well as of our understanding. The business of our highest intellectual faculties is, he says, 'to take thought of things noble and divine'. That, he says, certainly does not mean only the affairs of our own species, since he takes it as obvious that 'man is not the best thing in the world'.[7]

I strongly suspect that, in the end, some outward-looking, reverent attitude of this sort may be an unavoidable part of any serious pursuit of knowledge, and ought to figure in any explanation of its value. Mere intellectual predation –

71

fact-swallowing – simply is not enough to power effective thought. The world that we think *about* has to be seen as important, as having value in itself, if we want to claim that there is any great value in thinking about it.

Reflecting on this, I was struck by some remarks in Salman Rushdie's Herbert Read Memorial Lecture, 'Is Nothing Sacred?' This lecture was written shortly after the threats issued against him by Iranian clerics drove him into hiding, but well before he declared his reconversion to Islam. It seems to me to express well and honestly the dilemma that confronts a puzzled, undogmatic humanist today who really faces this inner conflict instead of merely externalizing it as a feud between various factions in the world. Rushdie wrote:

> It is important that we understand how profoundly we all feel the needs that religion, down the ages, has satisfied. I would suggest that these needs are of three types; firstly, the need to be given an articulation of our half-glimpsed knowledge of exaltation, of awe, of wonder; life is an awesome experience, and religion helps us understand why life so often makes us feel small, by telling us, what we are *smaller than*; and contrariwise, because we also have a sense of being special, of being chosen, religion helps us by telling us what we have been chosen by, and what for. Secondly, we need answers to the unanswerable; how did we get here? How did 'here' get here in the first place? Is this – this brief life – all there is? How can it be? What would be the point of that? And thirdly, we need codes to live by, 'rules for every damn thing'. The idea of God is at once a repository for our awestruck wonderment at life and an answer to the great questions of existence, and a rulebook too. The soul needs all these explanations, – not simply rational explanations, but explanations of the heart.[8]

Rushdie is surely right that, in framing our world-picture, it is essential that we should be able to *feel small* – to recognize our own unimportance. That means that we must acknowledge something else which we are *smaller than*. And this, after all, is not difficult. The immemorial human situation in the world has not been one of supremacy. People have always lived surrounded by living things and natural forces that were in many ways greater than themselves. They have responded to those things with awe and reverence, feelings which are not only the soil out of which religions have gradually developed, but are also a crucial part of our emotional and intellectual lives. If we claim that it is worth while to pursue physical science, we are surely committed to thinking that this awe and reverence are appropriate reactions to the physical world it studies. Otherwise, why bother?

SIGNIFICANCE OF THE FANTASIES

People like Monod, however, want us to get rid of all reverence, all belief in something greater than ourselves. This (I have been suggesting) does not actually result in outlawing all value judgment from thought about the physical world, because that is impossible. All thought about facts carries some value-judgment with it. Instead, this move replaces reverence by such feelings as contempt, horror, resentment, fear, hostility, estrangement and the ambition to dominate. It invites us to see the universe as something to be conquered, something beneath us, 'objective' in the sense of lifeless, drained of creativity and purpose, and it takes this to be the truly scientific attitude. The odd fantasies that I am citing in this book are expressions of that project. In compensation for the draining of value from the outside world, they orgiastically dramatize the human mind. They glorify *Homo sapiens* as the sole centre of value in a universe that exists merely to support him, and they ground that value primarily on a special use of the intellect, on the fact that human beings do science.

I am suggesting that these are not just casual pipe-dreams, not just symptoms of the way in which thought that is pressed down hard in one place tends to bulge out in another. The special choice of this particular place to press it down – the specific rejection of reverence, awe and sympathy for the world that we enquire about – is itself dangerously misleading. It misconceives the nature of curiosity.

What seems to me to emerge is this. If our curiosity is in no way respectful – if we don't see the objects we speculate about as joined with us and related to us, however distantly, within some vast enclosing common enterprise which gives them their independent importance – then (it appears) our curiosity, though it may remain intense, shrinks, corrupts and becomes just a form of predation. We then respond to these beings we enquire about with some more or less hostile, alienated attitude, something ranging between fear, aggression, callous contempt and violent suppression. We see them either as enemies to be conquered or as brute objects ranged over against us – as aliens, as monsters, as victims, as trivia or as meat to be eaten.

It is apparently not possible to take no attitude to them at all. Total neutrality, impersonality of the kind that has been recommended for the last three centuries simply does not seem to work. Projection is always present. People have never been able to be impersonal about Fate, and it seems it is not much easier to be so about Nature.

I do not, of course, mean that these sinister moods infect every moment of every study. We are dealing with one aspect of the motives for enquiry. My main business in this book is simply to show, by examples, that this is now quite an influential attitude, and to ask what it means. How much to worry about it, and what to do if one does worry, is then the reader's own choice.

7

PUTTING NATURE IN HER PLACE

WONDER – THE MECHANIST ATTACK

To return, then to the contemplative tradition – was Aristotle right to encourage wonder, awe and reverence towards the physical world? It is one of the points that the founders of modern science held against him. Thus Descartes wrote, 'Know that by nature I do not mean some goddess or some sort of imaginary power. I employ this word to signify matter itself.'[1] Similarly, Robert Boyle, in his *Enquiry into the Vulgarly Received Notion of Nature*, complained that 'men are taught and wont to attribute stupendous unaccountable effects to sympathy, antipathy, *fuga vacui*, substantial forms, and especially to a certain being . . . which they call Nature; for this is represented as a kind of goddess, whose power may be little less than boundless.'[2] Accordingly, Boyle complained, 'the veneration wherewith men are imbued for what they call nature, has been a discouraging impediment to the empire of man over the inferior creatures of God'.[3]

This was an important element in the new notion then being forged of what it was to be scientific. With a similar disapproval of wonder, Descartes earlier expressed the hope 'that those who have understood all that has been said in this treatise will, in future, *see nothing whose cause they cannot easily understand, nor anything that gives them any reason to marvel*'.[4] Wonder itself was to cease. Explanations were to become so clear that there was to be no more mystery. Not only would everything on earth now be understood, it would also be demythologized – disenchanted – depersonified and seen, in the bleakest of daylight, as not specially impressive after all. Matter, fully debunked, was from now on to be recognized as what the New Philosophy declared it to be – mere inert, passive, mindless stuff, devoid of spontaneity, of all interesting properties such as sympathy and antipathy, and above all destitute of any creative power. All pleasing forms that might seem to belong to matter were to be credited, not to it, but directly to God the Creator.

God, seen as fully active and fully intellectual, was the beneficiary now credited with these powers, reft from Nature. Having intelligence as well as creative power, God could do directly – either at the moment of creation or through later miracles – all that had been previously thought to need special adaptations in matter itself. And that is what the men who founded the Royal Society (by and large) took him to do.

It is surely extraordinary that nineteenth- and twentieth-century thinkers have supposed that they could take over this attitude to matter unaltered, while eliminating the omnipotent Creator who gave sense to it, as well as the immortal soul which took its status from him. The metaphor of matter as machinery still continues to run around like a chicken with its head off, though the Designer who gave a sense to it has been removed.

Peter Atkins, echoing Monod, rejoices that 'the Creator had absolutely no job to do' and 'can be allowed to evaporate and disappear from the scene'.[5] To make sure of that, it would be necessary both to understand much better what is involved in the idea of creation and to abolish the impoverished seventeenth-century ideas about mind and matter with a thoroughness that Atkins does not begin to conceive of. Before starting to raise any questions about a creating God, we need to make room for the creative powers of matter, to recognize once more the complexity of nature. The pre-adaptations that made life a possible option must, after all, still be lodged somewhere.

NATURE AND HER TORMENTORS

What went wrong? It may be easier to see that if we notice the way in which the pioneers of mechanism went about reshaping the concept of Nature. Very properly, they wanted to try the experiment of depersonalizing it. With that in view, the first step they surely needed to take was to stop using the feminine pronoun, or indeed any personal pronoun for 'Nature' altogether. But this was not done. We come here to one more of the strange compensatory myths, dreams or dramas that are my theme. The literature of early modern science is a mine of highly-coloured passages that describe Nature, by no means as a neutral object, but as a seductive but troublesome female, to be unrelentingly pursued, sought out, fought against, chased into her inmost sanctuaries, prevented from escaping, persistently courted, wooed, harried, vexed, tormented, unveiled, unrobed, and 'put to the question' (i.e. interrogated under torture), forced to confess 'all that lay in her most intimate recesses', her 'beautiful bosom' must be laid bare, she must be held down and finally 'penetrated', 'pierced' and 'vanquished' (words which constantly recur).

Now this odd talk does not come just from a few exceptionally uninhibited writers. It has not been invented by modern feminists. It is the common, constant idiom of the age. Since historians began to notice it, they have been able to collect it up easily in handfuls for every discussion. I can't spend time on doing that here, but I will just give briefly a few well-known examples from Francis Bacon, who was something of a trail-blazer in the matter.

Bacon dismissed the Aristotelians as people who had stood impotent before Nature, destined 'never to lay hold of her and capture her'. Aristotle (said Bacon), being a mere contemplative, had 'left Nature herself untouched and inviolate'. By contrast, Bacon called upon the 'true sons of knowledge' to 'penetrate further' and to 'overcome Nature in action', so that 'passing by the outer courts of nature, which many have trodden, we may find a way at length into her inner chambers'. Mankind would then be able, not just to 'exert a gentle guidance over Nature's course', but to 'conquer and subdue nature, to shake her to her foundations' and to 'discover the secrets still locked in Nature's bosom'. Men (Bacon added) ought to make peace among themselves so as to turn 'with united forces against the Nature of Things, to storm and occupy her castles and strongholds'. By these means scientists would bring about the 'truly masculine birth of time' by which they would subdue 'Nature with all her children, to bind her to your service and make her your slave'.[6]

Just to show that this way of talking did not die with

the crude manners of the seventeenth century, here are a couple of later echoes from Adam Sedgwick, that immensely respectable clerical professor of geology at Cambridge who was so disturbed by Darwin's theories. Sedgwick, describing true scientific method, explained how, after laws have been carefully formulated, investigators must always 'again put nature to the torture and wring new secrets from her'.[7] And, shifting to the military end of the spectrum, Sedgwick also described Newton as having 'stormed the sky with mathematical artillery'.[8]

THE CLEANSING FIRE

As I say, these quotations are not exceptional. If we were just looking for absurdities, and trying to show the failure of the impersonal stance, we could spend many instructive hours sifting a crowd of still more picturesque examples. But the point is, of course, not just to collect them but to understand what is going on. In real life, most of these distinguished scholars were neither sex-maniacs nor soldiers sacking a city. Most of them, probably, would not normally have hurt a fly. Why, then, did they continually use this kind of language? Three explanations suggest themselves, one to be rejected, two to be seriously considered.

1 (Negative) They were *not* just exceptionally naïve. All ages, including our own, are naive in their own way. Past errors only differ from present ones in being easier to see.
2 (Positive) They were trying to develop the very peculiar idea of matter as wholly inert, passive and unproductive, without any spontaneity or interesting qualities. This idea was far more entangled in traditional gender symbolism than they realized, because earlier, Aristotelian science – most bizarrely – deemed women also to be essentially inert, passive and unproductive, mere vehicles for reproduction. As a piece of science, this notion of matter has gradually been shown up as inadequate and misleading by the later developments in physics. But as a drama, it has had enormous power, and derivatives of it still have a strong confusing influence, both in scientific and in everyday thinking. They are involved in most of the strange later fantasies we shall be looking at. So this is a point which still concerns us.

3 (Also positive) Besides this unsatisfactory doctrine about matter itself, and the gender symbolism, further trouble was introduced by the destructive gusto that, from the start, went with it. Wanting to emphasize experiment, the pioneers of modern science had an image of themselves which differed from most earlier images of learning in being more workmanlike, more suggestive of physical violence. This physicality, together with the fact that they really did want to make big changes, led them to revel in drastic language.

No doubt scholars proposing new schemes always have slashed at existing ones. But there really was a crucial shift of emphasis in the early Enlightenment towards making this destructive cutting and slashing central, and towards seeing the gusto that goes with it as a central motive for science. It began to seem that a scientist is typically a destroyer, one who sweeps away existing superstitions, rather than one who works to construct further on existing foundations. And among these superstitions, the former idea of Nature seemed an obvious target.

The difficulty about this destructive approach is of course how to keep some discrimination about what to destroy. Not all destruction is helpful. Almost any destructive move involves a positive one as well, and the gratifying sense that one has killed something bad can distract attention from the details of what one is promoting instead. Mechanistic seventeenth-century scientists displayed a new purifying zeal, a passion for disinfection, at times a cognitive washing-compulsion, accompanied by a rather touching willingness to accept even a minor role in the great cleansing process. And these too came to be seen as essential to science.

The impersonality aimed at in modern science did indeed find its place here. Bacon said that experimental philosophy 'goes far to level men's wits'[9] because it 'performs everything by surest rules and demonstrations'. Since anyone can scrub, scientists might offer themselves as humble fellow-workers without seeming to assume any pretentious role reminiscent of earlier sages, and without being held responsible for the main planning of the building. Thus Henry Power, celebrating the Royal Society in 1664, cried out,

> Methinks I see how all the old Rubbish must be thrown away, and the rotten Buildings be overthrown, and carried

away with so powerful an inundation. These are the days that must lay a new Foundation of a more magnificent Philosophy, never to be overthrown . . . a true and permanent Philosophy.[10]

John Locke showed the same spring-cleaning spirit in the famous Introduction to his *Essay on the Human Understanding*:

The commonwealth of learning is not at this time without master-builders, whose mighty designs in advancing the sciences will leave lasting monuments to the admiration of posterity. Everyone must not hope to be a Boyle or a Sydenham, and in an age that produces such masters as the great Huygenius and the incomparable Mr Newton, with some other of that strain, it is ambition enough to be employed as an under-labourer in clearing the ground a little, and removing some of the rubbish that lies in the way of knowledge.[11]

It is tempting to set Locke's lively picture beside a less cheerful view of the academic building-site, as Conrad Waddington saw it three centuries later:

Scientists have tended to refuse to see the wood for the trees. There have been an army of bricklayers piling brick on brick, even plumbers setting up super WCs, and heating and lighting engineers installing the most modern equipment; but they have all united to shoo the architect off the building site, and the edifice of knowledge is growing like a factory with a furnace too big for its boilers, its precision tools installed in a room with no lighting, and anyhow with no one who knows what it is supposed to manufacture.[12]

This situation might, of course, have something to do with the modest tendency, which Locke praised, to leave other people to take the big decisions, including the decision about what counts as rubbish to be carted away. Unfortunately, as things turned out, none of the great architects Locke named did come up with a comprehensive plan for science. And there is growing evidence that Newton, had he been asked to do so, would have produced a plan centring the edifice on alchemy . . . In any case, however, it is doubtful policy to romanticize the destructive emphasis as

Locke and his friends did – to cultivate the pugnacious zest that accompanies a release from positive choice.

SCIENTISTS EMBATTLED

What, then, was all this destructiveness directed against? It is evident that, at this point, there did develop a sense of real alarm and disgust – a resolve to *écraser l'infâme* – directed against earlier views which were seen, not just as mistaken, but as odious because religiously wrong – as pagan and superstitious.

The campaign waged by members of the Royal Society, and by seventeenth-century mechanists generally, was not, as their atheistical successors often suppose, a campaign against religion as such. It was primarily a campaign against the *wrong* religion – against what seemed like nature-worship, against a religion centring on the earth, and apparently acknowledging a mysterious pagan goddess rather than an intellectual god. All the great scientific pioneers claimed to be campaigning on behalf of Christianity. And with most of them this was not just a political move – as again people now tend to think – but a matter of real conviction.

Nor was the fight only against Aristotelian thinking. Aristotelianism was indeed the traditional orthodoxy that all scientific reformers wanted to change. But the contest was three-cornered, and the most bitter hostility was between two parties of reformers – between the mechanists, represented by Descartes, and the exponents of what was called 'natural magic'. This was a belief in an all-pervading system of occult forces, of mysterious sympathies and antipathies, the sort of thing that we do indeed now tend to think of as superstitious.

It was, however, by no means just a hole-and-corner affair used by sorcerers. It was a sophisticated system expounded by scientists some of whom were not in any ordinary sense magicians at all, but were quite as learned, quite as experimental, and often quite as successful, as the mechanists. The contest was not a simple one between light and darkness.

Thus, Galileo in his *Dialogue* wrote with great respect of William Gilbert's book *De Magnete* (1600), accepting Gilbert's findings about magnets, though he differed from him about how to interpret them. Gilbert had attacked Aristotle for dividing the cosmos into a divine realm in the heavens and an inferior one on the earth, because this view dishonoured the earth. The earth, wrote Gilbert, is not to be 'condemned and driven into exile and cast out of all the fair order of the glorious universe, as being brute and soulless'. 'As for us,' he continues, 'we deem the whole world animate and all globes, all stars, and this glorious earth too, we hold to be from the beginning by their own destinate souls governed.'[13]

Gilbert and Galileo thus both wanted to bring attitudes to the earth and heavens together again. But Galileo saw this as best achieved by withdrawing superstitious reverence from the heavens while exalting the earth. 'We shall prove the earth to be a wandering body surpassing the moon in splendour, and not the sink of all dull refuse of the universe.'[14] Gilbert, by contrast, proposed to do it by extending reverence to earth as well as heaven, by looking for explanations of its behaviour in its own creative properties, and by the very significant image of the earth as mother. Gilbert wrote that all material things have 'a propensity . . . towards a common source, towards the mother where they were begotten'.[15]

In some contexts, these ideas proved surprisingly useful for science. For instance, Gilbert argued that tides are produced by the attraction of the moon, working through sympathy. Johannes Kepler, accepting this idea, added that this was only part of a general system of attraction which explains all 'heaviness (or gravity)'. Heaviness, said Kepler, is simply a 'mutual corporeal disposition between related bodies towards union or conjunction . . . so that it is much rather the case that the earth attracts a stone than that the stone seeks the earth'. Kepler suggested that the moon's attraction is what produces the tides, and he added, 'If the earth should cease to attract its oceans, the waters in all its seas would fly up and flow round the body of the moon.'[16] Kepler built this idea into his refinement of the Copernican system, by which he produced tables of the planetary motions which were some fifty to a hundred times more accurate than existing tables.

To us, who are used to Newton, all this seems reasonable enough, and Kepler may sound like a typical pioneer of modern science. But this is where our foundation-myths are so misleading. At the time, the mechanistic scientists who fill the rest of our pantheon rejected Kepler's view fiercely as superstitious. In particular Galileo, who might have been expected to welcome Kepler's support for the Copernican system, simply ignored it. The trouble was that, in the mechanists' view, 'attraction' was no real explanation at all. It was just an unintelligible, vacuous name for an 'occult force'.

To mechanists, no explanation counted as intelligible unless it worked on the familiar model of push-pull, like the parts of the cog-driven machines with which they were familiar. Now it is hopelessly difficult to explain in this kind of way the well-known fact that things fall, or indeed how things stick together in the first place – how the hard particles, whose motion leads them only to bang against each other, sometimes form solid stones rather than heaps of dust. Attraction was suggested here too, but it was still viewed as a vacuous superstition.

The mechanistic systems most widely favoured, such as Descartes's theory of vortices, had no explanation for either of these things that looked even faintly plausible. In spite of this, not only was Kepler laughed out of court, but the same objection still told very strongly later against Newton. His theory of gravitation was resisted as empty and irrational well into the eighteenth century. As late as 1747, three most distinguished French scientists – Euler, Clairaut and d'Alembert – claimed to have disproved Newton's theory of gravitation, and it was some time before the resulting controversy was settled in his favour.

THE REMARKABLE
MASCULINE BIRTH OF TIME

EXPLANATIONS AND RATIONALITY

All this lights up in a most interesting way the question of *what counts as an explanation*. Familiarity is always demanded here. The mechanists thought it more rational to stick with patent non-explanations – with stories that did not pretend to explain at all – rather than to use an explanation that was fertile but unfamiliar in form. They thought, moreover, that rationality demanded complete simplicity; there must be only one explanatory system. They should therefore leave what was effectively a blank round awkward facts such as the phenomenon of falling bodies, until they could explain it by a story of the only right and familiar form.

Their faith that this better story would follow is impressive. Descartes laid it down as a demand of reason that the post-dated cheques would, in the end, always be honoured. It was certain that all temporarily puzzling items – such as magnets, tides and falling bodies – have 'no qualities so occult nor effects of sympathy and antipathy so marvellous or strange', that their properties cannot be explained in terms of the *'size, shape, situation and motion of different particles of matter'*.[1] (The differences among these 'different particles' themselves would, of course, only be differences of size and shape; all matter was otherwise homogeneous, inert and without qualities.) Again, 'There are no amazing and marvellous sympathies and antipathies, in fact *there exists nothing in the whole of nature which cannot be explained in terms of purely corporeal causes devoid of mind and thought.'*[2]

It is interesting to see how Descartes's double negative here conceals the huge confidence of his claim. To say 'there is nothing that cannot be so explained', sounds quite sceptical and parsimonious. To say, 'I and my colleagues can and eventually will explain everything in these limited terms', would sound much bolder. But they come to the same thing. The claim is that, in the end, Nature will be forced to speak the whole truth in this one language. If it seems to be saying something in some other language meanwhile, it should not be listened to.

This ruling went for gravitational attraction, and it also went for the even tougher case of living creatures. No special creative properties of matter – no *biological* properties – were to be allowed for the forming of these. Descartes declared them to be automata, mechanisms working by arrangements of inner cogs and pistons which were not even very complex. The physicist's chronic lack of interest in biology has seldom been so plainly expressed.

'Since' (wrote Descartes) '*so little is necessary to make an animal*, it is certainly not surprising that so many animals, so many worms and insects, are formed spontaneously under our very eyes in all kinds of putrefying matter'[3] (Emphasis mine). These animals had of course no souls and were no more conscious than the rest of the physical world. The size, shape, situation and motion of particles would easily explain them. Or, as Peter Atkins put it in the quite recent formulation that I quoted earlier, 'Inanimate things are innately simple. That is one more step along the path to the view that animate things, being innately inanimate, are innately simple too.'[4]

The striking thing about claims like these is surely the high proportion of faith to evidence. The force supporting this faith is not any observation of facts. It is a special, very narrow, picture of what scientific rationality demands, a picture which allows only a small set of premises. It is assumed that all explanations will be of one type, that they must all be expressed in a single language.

This assumption did of course bear good fruit where mathematics was taken to be that language, by making it possible to discover general formal structures underlying matter. But one successful set of explanations never rules out the scope for others. It is not possible that mathematics itself should do all the explaining that we need. In order to apply mathematics to the real world at all, we have to use other conceptual schemes first so as to select the items that are to be measured or counted. Just as mathematical expressions are only a small, specialized part of the ordinary language we speak, so mathematical explanations are evidently only one small part of the range of concepts by which we explain and understand things.

What seventeenth-century rationalists like Descartes hoped to do was to build round mathematics a single system of concepts continuous with it, which would be uniform, and able to give a unified explanation of the physical world. It was a magnificent idea, and nobody could know whether it would work till it had been tried. But its champions did not just try it; they declared for a fact that the world was such as to make it work. What was the basis of this faith? Brian Easlea comments:

> The mechanical philosophy certainly presents a breath-taking conception of matter and the cosmos! To say the least, its truth does not stare the natural philosopher in the face. Why, then, was it so widely subscribed to? It was one thing to reject a powerful, creative 'mother earth'; it was quite another to declare nature to consist only of inert, uninteresting matter and nothing more! . . . On the credit side, it was at least a transparently clear philosophy; *matter became, perhaps for the first time and undoubtedly for the last time, conceptually graspable by natural philosophers.* Nevertheless, despite the undoubted advantages of conceptual clarity, the proponents of the mechanical philosophy experienced the utmost difficulty in satisfactorily accounting for such ubiquitous phenomena as cohesion and the falling of heavy bodies perpendicularly to the earth's surface, not to mention the nature of the mind's interaction with matter, how spontaneous generation occurs and how embryos are formed. (Emphasis mine)

Easlea goes on to make – what is surely called for – a suggestion about what the extra, non-scientific motives might be that caused this very unsatisfactory piece of science to gain such authority:

> What the mechanical philosophy amounted to was, it seems, a radical 'de-mothering' of nature and the earth in preparation for, and legitimation of, the technological appropriation of the natural world that the mechanical philosophers hoped they and their successors would undertake.[5]

That is to say, the success of this approach was not due – as we were brought up to believe – solely and directly to its scientific correctness. Much of its science did turn out brilliantly successful and, on present views, correct. But the success was quite uneven. Other parts of it were, on present views, simply wrong, and were felt even at the time to be inadequate, though they sprang directly from motives which were conceived as scientific. Some problems – notably that of the connexion between soul and body – were admitted to be so awkward for the mechanistic approach that they could only be solved by assuming a perpetual miracle. The need for this miracle was then welcomed as proof of the existence of God, who was needed to perform it.

If we ask how this kind of explanation by miracle differs from the assumption of occult forces, the answer plainly cannot be that it is intellectually clearer. It must be that the religious doctrines involved are sounder. And for some of these awkward problems – such as gravitation – other, non-mechanistic scientific schools had better explanations available.

THE TEMPER OF THE AGE

Of course the solid scientific achievements did play a great part in ensuring the success of the mechanistic approach. But they were supported by something much deeper and less clearly recognized – by a temper, a mood, a drama that filled a felt need at the time, and has long continued to fill it.

Mention of this does not need to tip us into an unbridled relativism. You don't need to be a full-time Marxist to see the economic attraction of a free licence to exploit nature for the age that was beginning to feel the stirrings of the Industrial Revolution and of colonial expansion. You don't need to be a full-time Jungian to see that the symbolism of Mother Earth is a strong one, which can seem threatening to people who are struggling to establish their own independent identity. The unacknowledged Anima can take some very alarming forms, and this alarm can generate bitter and destructive resentment. The denied female element within the male character was clearly giving trouble. But so, surely, were actual women in the world.

For I think you don't – finally – need to be a full-time feminist to conclude something more. When a school of thought, officially dedicated to clear, literal, unemotive speech, regularly uses a lurid language of sexual pursuit, torture and rape to describe the interaction between scientists and the natural world, trouble is also surfacing about the relations between actual men and women. At such a point, an entry in the index under the heading 'gender insecurity' doesn't seem excessive.[6]

The mention of feminism in these discussions tends to produce a charge of irrelevance; why talk about that? But it is not irrelevant to correct a long-standing bias. Virism really has been chronic, and has produced some surprisingly irrational distortions. It has not been just a promotion of men's interests over women's but an obsession with a distorted ideal of maleness, an ideal which can in fact damage men's lives as well as women's.

When modern science was being formed, some consciousness of trouble about this sex-linked ideal was already arising. As we are all told at school, the Renaissance was the age of dawning individualism. It was the time when ancient hierarchies began to break up, when kings had their heads cut off and wars of religion subverted societies. It was a time of great insecurity, in which the promise of order which science offered was welcomed for other reasons besides intellectual ones. It was also, however, the age when print diffused learning far more widely than ever before, so that people heard of conditions other than their own, and became less willing to accept subjection. Already in the seventeenth century, some women (especially in France) were beginning to get a little of that learning, and beginning also to become a nuisance by asking for more. We must surely notice that this ingredient too went into the pot if we want to account fully for the strange stew that came out of it.

In any case, what begins to emerge is that the debunking of matter, the desacralizing of the earth, did strike a chord in many people that made it plausibly appear as their salvation. The particular danger that they were struggling to be free of seemed like subservience to an irrational queen or mother. That is why they welcomed – in Bacon's extraordinary phrase – the prospect of a 'masculine birth of time'. That was why the spokesmen of the Royal Society repeatedly declared that it existed to promote 'a truly masculine philosophy'. (What could that possibly mean?) That was why there was such an outbreak of bizarre sexual metaphors in writings about science – an outbreak which (I repeat) is quite real and not an invention of modern feminists. And that is surely also why the simple, clockwork machine model, in spite of its startling faults, kept so much prestige and has remained popular for so long, even when it has proved unusable in many areas, notably in particle physics.

THE TRIUMPH OF FAITH

That, too, was surely why confidence in science as such was so euphoric; why post-dated cheques for the future were so willingly accepted. Descartes's complacency is not exceptional here. From his time on, it has repeatedly been firmly claimed, both that particular scientific programs will soon produce complete explanations and that science as a whole is about to do so. Indeed, the belief that it will is, again, a part of what has been held to be a scientific attitude.

Over the particular programs, these claims have again and again proved delusive. They usually turn out to be merely the effect of the over-confidence that comes over hard-working people when some success does at last reward them. About science as a whole, many distinguished sages have pointed out that these claims are bound to collapse. All answers raise more questions; all explanations are provisional and incomplete. Yet the claims go on. Lord Kelvin's declaration towards the end of the nineteenth century that physics was virtually complete was only one of them. Much more recently, Peter Atkins (1981) went on record as follows:

> When we have dealt with the values of the fundamental constants by seeing that they are unavoidably so, and have dismissed them as irrelevant, *we shall have arrived at complete understanding.* Fundamental science can then rest. *We are almost there. Complete knowledge is just within our grasp. Comprehension is moving across the face of the earth, like a sunrise.*[7] (Emphases mine)

Similarly Stephen Hawking seems to hope that a complete cosmological theory can be produced which will make possible 'the ultimate triumph of human reason', namely that 'we would know the mind of God'.[8]

It is worth while to remember this kind of remark when we come across the frequently held opinion that hard-headed incredulity is a central part of the scientific character. For scientists, as for anybody else, incredulity is bound to be selective. The wholesale commitment of seventeenth-century physicists to their models was – like that of Darwin, mentioned earlier – for their time an invaluable means of getting the best out of those models. But that is no reason for taking literally the claims that expressed it . Claims like these are chiefly interesting as proofs of what I have called a faith. They have, I think, very little to do with their official subject-matter – with any real question about the content and prospects of science itself. The commitment is always to the drama. What then is that drama?

THE CONFRONTATION

It presented itself to the seventeenth century as a conflict between light and darkness, but also between male and female as alternative possible creative principles. The female principle – Nature, as conceived in the doctrine of natural magic – was life-giving, fertile, bountiful and generous, but also dark in the sense of mysterious and vast. Because mystery may always conceal danger, she might well be dark also in the sense of sinister and threatening. By contrast the masculine principle – the divine Creator – supplied order for that life, but produced light as well – light which is essential for life as well as for understanding.

Could not these two elements have been seen as co-existing and co-operating? Was it necessary to choose between them? In earlier times it had seemed more or less possible to combine them, though the attempt to do so often produced serious tension. But in the seventeenth century, most disturbingly, it somehow became much harder to bridge this gap. The male and female principles increasingly appeared as alternatives, indeed as opponents.

The reason for this alarming breach looked different from different angles. From an intellectual angle it appeared, quite respectably, as simply a wider curiosity, an intenser thirst for knowledge. Reason (it seemed) had raised its standards. It was no longer content with a limited understanding of the physical world. It demanded to penetrate everywhere. For this purpose the light must drive back all the darkness; there must be no more mysteries left. From the religious angle, the project again seemed straightforward and honourable; it was simply a matter of doing due service to God the Creator by making clear that he was responsible for everything of value in his Creation.

Why, however, should either of these harmless projects have called for violent gender-imagery, or indeed for gender-imagery at all? The answer, unfortunately, seems to be that this kind of symbolism has far deeper roots than may appear. It is not a casual weed but a structural feature of the social landscape. Even writers who notice its distorting effect often flounder instructively in their efforts to correct it. Thus Robert Hooke likened matter to the 'Female or Mother Principle' which was (he explained) 'abstractly considered without Life or Motion, without form, and void, and dark, *a power in itself wholly unactive*, until it be, as it were, impregnated by the second Principle, which may represent the Pater, and may be called *Paternus, Spiritus*, or hylarchic Spirit'.[9]

This way of thinking about male and female was of course not new. As already mentioned, it, with its accompanying physiological fairy-tales, goes back at least to Aristotle, and had been accepted throughout the Middle Ages. But we surely have a right to ask why the bold, revolutionary iconoclasts of the seventeenth century could not throw out this piece of intellectual garbage along with so much else. And we are forced, I think, to give the obvious answer. From the psychological angle, male domination had always been insecure and uneasy. In an age of political revolt and increasing individualism, that domination felt less secure than ever. It was no longer enough for the rational male principle to be seen as the source of all order. He must now be altogether omnipotent. He must control and monopolize all the sources of life as well.

91

9

UNEXPECTED DIFFICULTIES OF DEICIDE

PURIFYING THE CONCEPTUAL BACKGROUND

I have been suggesting that the idea of science necessarily has a much wider function in our lives than the neutral one of merely purveying information about a world conceived as alien and 'objective'. In so far as it is serious, its wider outlines express a world-picture that deeply concerns us, that shapes the meaning of our life, that affects our salvation. And I have backed that view by pointing out the dramas that have surrounded the arguments of thinkers officially promoting a neutral approach, such as Jacques Monod and the seventeenth-century pioneers of modern science.

It may naturally still seem, however, that these defects only show that they did not try hard enough. Are the confusions we inherit from the scientific pioneers of the seventeenth century just distressing symptoms of weak will, proving once more the necessity of determined atheism? Would this kind of rhetoric vanish if only the scorched-earth policy were completed?

That is the diagnosis that has been widely accepted. Atheists from Hobbes to Monod and Sartre have been sure they had the answer. With stronger will-power, sterner resolve, firmer defiance of society, the supernatural could finally go and would never be missed. The murder must be firmly done and the body disposed of. A proper concept of matter could be reached by a purgative process – by scouring away the alien ideas which obscured it – thus perfecting a truly scientific outlook.

This drastic approach is the mirror-image of the one that Plato used from the other end, when he tried to clarify his concept of the soul. He wrote:

To understand her real nature, we must look at her, not as we see her now, marred by association with the body and other evils, but when she has regained *that pure condition which the eye of reason can discern* . . . Our description of the soul is true of her present appearance, but we have seen her afflicted by countless evils, like the sea-god Glaucus whose original form can hardly be discerned, because parts of his body have been broken off or crushed and altogether marred by the waves, and the clinging overgrowth of rock and shell has made him more like some monster than his natural self.

The soul therefore (said Plato) cannot be properly understood till it has obeyed

the impulse that would disencumber it of all that wild profusion of rock and shell, whose earthy substance has encrusted her, because she seeks what men call happiness by making earth her food.[1]

WORLD-PICTURES AND INNER CONFLICT

Just so, according to campaigning materialists such as Hobbes and Laplace, the concept of the physical world had been obscured by being long immersed in superstition. It must be scrubbed free from its animist accretions, from all talk of God and the soul, so that it can regain 'that pure condition which the eye of reason can discern'. This was actually a far more drastic metaphysical plan than Plato's. Plato did not try to get rid of the physical world. He thought that, in its own way, it was real; he simply considered it foreign to the soul and therefore dangerous. But the materialists did want to get rid of the spirit.

Both parties were surely engaged in premature, vicious abstraction. The first point to consider is not whether there is a God, nor whether there is a break at death between an immortal soul and a mortal body. It is how, in this life, we are to view and interpret both the world around us and the world within us. We need ways of thinking which are unifying enough to give us guiding patterns, but not so strongly reductive as to leave out something important.

It is quite hard to frame these patterns, and our choice among them is commonly a way of taking sides in some inner conflict. Campaigning animists and materialists are both, first and foremost, expressing an allegiance to a way of living, and only secondarily talking about the world or our prospects after death. Thus, Plato was dedicated to the contemplative life, and Enlightenment atheists were aiming at an existence free from the political and social control of the Church.

It is scarcely possible, however, for this kind of allegiance to be single-minded, because the choices it offers are too narrow. Taking sides about metaphysics often requires us, not just to ignore some data, but also to suppress some part of our personalities. If this suppression is done openly, as an explicit moral choice, it can be legitimate. We do, after all, have to sacrifice some aims to others in a limited life. But if we do it unconsciously, under the guise of merely recognizing a fact about the world, we make grave trouble.

Both campaigning animists such as Plato and campaigning materialists in the tradition of Hobbes have done this. Both have ignored the complexity of human personality by drawing a sharp line round what could legitimately be said to belong to it, to suit their over-simple purposes. Both, I think, have uneasily suspected that this might be so, and their suspicion accounts for an irrational, fanatical ferocity in their respective scouring programs. In Plato's case, this has been often noticed. In the materialists' case it has not, because until lately we have been living more or less inside their program. But it needs to be.

Removing spirit from the Cartesian system makes serious trouble about what is then left. The concept of the natural world was originally tailored to fit the current supernatural one. It does not make sense on its own. In some ways, it is only a shadow of its supernatural partner. Life and vigour have been deliberately drained from it, as well as from the rest of the spiritual realm, to be conferred on the omnipotent Creator. All the eggs were thus put in one basket.

This made removing that Creator – and the soul that was his representative – a most dangerous and difficult psychological operation. He carried a huge freight of meaning which was not at all fully understood. In the Enlightenment, however, many reformers thought this task quite simple. It appeared as just the final move in a long process of simplistic controversy and

94

heresy-hunting within the Christian tradition itself. To deny that God and the soul existed seemed only the logical next move after denying doctrines such as the Trinity or transubstantiation or the efficacy of prayers for the dead. And all these denials appeared, in an important way, like denying that there were unicorns or that witches could kill by cursing. They all seemed to concern matters of fact, determinable by evidence.

UNICORNS AND RHINOCEROSES

By this method only two alternatives are considered. There is, or there is not, a unicorn in the garden. If there is not, then there is nothing there at all. The rhinoceros or antelope that may be there is of no interest, no matter for surprise or wonder. Nor are the flowers, the trees or the soil. It is not guessed that they might now need to be looked at differently. If a unicornless world proves to be one drained of significance, then it is concluded that the significance, as much as the unicorns, always was a mistake.

Without significance, however, people cannot live. To see life as having a meaning is not just to add an indulgence, a colour or a taste, to its raw data. It is to find any shape in it at all, any connexion among its elements. This is not a luxury; it is the condition which makes thinking possible. The question is not whether we are pro- or anti-God. It is: how do we now map the connexions in the world if they are not to be described by talk of God? What sort of world do we now have? Connexion itself is not a superstition that we can get rid of. It is work that must be done one way or another. To refuse that work will not stop it being done. It will only leave it to the uncontrolled play of the imagination.

Failure to see this complexity is not a new fault, invented by the modern world. It is a batch of ancient faults taken over unnoticed from the Christian tradition, or, more exactly, from its entanglement in political feuds, which committed it to constant polarization about simple dogmas. In the seventeenth-century wars of religion, as in earlier disputes, enormous issues of doctrine were repeatedly treated as factual questions with a single right answer, reachable through controversy.

Once political sides had been taken, it became extremely hard to suggest that the truth is so vast that both these doctrines may be only attempts to grasp at a part of it. Instead, nations confidently drilled their peoples to accept <u>one of two solutions,</u> while dissenters, just as confidently, died proclaiming the other. With the same sort of confidence, atheists now pronounced their own final solution. As a matter of simple fact, they explained, there was no God, and – equally as a matter of fact – the physical world was (by sheer good luck) orderly, constructed just as it needed to be for scientific enquiry every bit as well as if God had done it.

MEGAFACTS AND METAPHYSICS

Vast propositions like these, however, are not very like everyday matters of fact. Are they matters of fact at all? What does it mean to call them so? What is the alternative? Current usage thinks only of 'value-judgments' which is far too narrow. Very general statements about the way the universe works, such as that it is ordered, or is – as Monod claims – totally contingent, or that it is, or is not, an illusion, or that it is in the hand of God, or that all events in it are causally determined, or that it is only a social construction, are not judgments of value. Least of all are they unaccountable judgments of value, of the vague kind which people often seem now to mean by that term. They certainly do not just say 'boo' or 'hurray'.

What they have in common with ordinary, modest factual statements is that they are intended to be true or false – to describe some actual state of affairs, not to be fiction. Where they differ is in that it is much less obvious how we can know them. We cannot compare them directly with any actual thing or things; they are far too wide. We cannot test them, as we do reports about unicorns, by the ordinary rules of evidence, relating such reports to a batch of neighbouring facts. There are simply too many facts involved.

The whole world cannot be brought before us and checked over to show that it is entirely orderly, or entirely contingent, or entirely determined, still less to show whether it is illusory or whether any reality underlies our social constructions. What we can check by experience is always only a tiny fragment of what we need to believe. Our world-pictures are vast imaginative extensions raying out from that experience. They are not drawn at random, but generated by our imaginations on such principles as they find natural and helpful for the sort of understanding

96

we need. The assumption of orderliness in the world is one of those principles. It is not a conclusion of science; it is something assumed in order that science can be done. When we fail to find order, scientific method itself demands that we still have faith that it is there.

We are not free (it must be repeated) to make up these assumptions just as we fancy. Only within quite narrow limits can they be called 'social constructions'. A world-picture without any order would be useless, and the order must be such as to mesh with the central structures of experience. No society can construct a world in which falling off cliffs will not hurt, nor even one in which people do not have inner conflicts. And again, the idea that everything we think about is a social construction could not itself be proved as a fact. Yet such general ideas are certainly not offered as being themselves just one more convenient social construction. They are argued for as being true.

[margin note: Order vs. disorder]

THE NEED FOR INTERPRETATION

[margin note: matters of fact vs. matters of value]

We are dealing here with a wide realm of general thought that lies between the two small enclosures now called matters-of-fact and matters-of-value. Some of this realm consists of rather general theories about particular kinds of facts. Some of it deals with matters more general still, and is called metaphysics.

The first part includes scientific theories, for instance about such things as continental drift, or the Big Bang, or the origin of species by natural selection. These theories do appeal to particular, experienced facts as evidence. But their main function is not to record these facts, but to interpret them in a way that will connect up other known facts in a pattern that makes them understandable. If it does, it is held to justify belief in further facts that are not yet known.

These theories often allow us to infer amazingly large fields of further unexperienced facts. Science does not mind this, provided that the reasoning is logical and there are not too many other data contradicting them. In fact, enquirers sometimes accept even theories which contradict quite a lot of the data, in the hope that something will turn up later to explain the clash.

[margin note: Yes!]

As is now well understood, scientific thought thus does *not* build its theories on the old, inductive model, by collecting data

97

first and only generalizing them into a theory when enough of them are in complete agreement. The proportion of data to new facts inferred is often quite low. In the case of the Big Bang, for instance, immense facts are boldly inferred from relatively slight data by the use of very elaborate, far-reaching conceptual schemes which are held to have proved their validity. Science, in fact, is not at all the simple direct record of experience which extreme empiricist theory once supposed it to be. Neither is history, nor any other branch of learning. They all consist largely of conceptual schemes devised to make understanding possible.

Really?. then what is it?

Once we grasp this, it should be clear that there is nothing specially fishy about the yet wider thinking which is metaphysics. Here we use still bigger conceptual schemes, concerned not just with particular batches of facts but with the entire world, with facts as a whole. Some of the ideas suggested here are weird and startling, such as that the whole world is an illusion, or that it is radically disordered. But the contraries of these views – ideas which sound most sensible and obvious – are part of metaphysics too. They pose just the same problem about how they can be known. Of course since we find them useful, we do not usually bother about that problem. Raising it tends to upset us, which is why metaphysics has such a bad name.

THE ATTEMPT TO AVOID METAPHYSICS

During the early twentieth century, great efforts were made to dodge this problem by saying that big metaphysical statements like this did not need to be known, because they were actually meaningless. What difference can it make (theorists asked) to say that everything is an illusion, or that everything is contingent, so long as this applies to everything equally? If my toothache is just as illusory as my unicorn, and a plain causal sequence is as contingent as a casino, then the standard of illusion or contingency has been destroyed, and these words have lost their meaning. Accordingly, all metaphysics was held to be simply vacuous talk, and the word 'metaphysics' itself was used as a mere term of abuse. (This habit, which is now merely illiterate, still sometimes persists.)[2]

There is, of course, a real point here. It is true that metaphysical remarks do not have an obvious and complete sense, already given independently of their context. Anyone who makes them needs to give them a sense, to make their bearing clear, to say why they have arisen, and especially to say what is meant to follow from them. It is always reasonable to ask 'What do you mean by that?' about metaphysical remarks in a way that would not be reasonable if somebody said 'Breakfast is ready', or 'Get the fire-brigade'. Someone who says that all is illusion, and when asked to explain merely answers, 'Oh I don't know; it just is', and adds no more, does indeed seem not to be saying anything at all. A context is needed.

Metaphysical remarks need context [handwritten marginal note]

But it is not only metaphysical remarks that need a context. Most talk needs one, and the wider its reference, the more badly it needs it. The reason why people are often baffled and maddened by metaphysics is that they do not see why these vast things are being said at all. They do not see what is being *denied* – what previous widespread trouble with the conceptual drains caused the metaphysical plumber to be sent for in the first place. It can become particularly hard to see this after he has done his work successfully and cured the error. But it is also sometimes hard to see the error because it has not yet been cured – because we are still living with the bad smell, and are so used to it that it never occurs to us to want it removed.

PRACTICAL IMPLICATIONS

Metaphysicians are not just prophets. If they know their business, they do not simply throw out remarks without a practical context. They give reasons why we ought to change our view of the world. These need to be solid reasons, arising out of the rest of our thinking, out of the logical need to be consistent with other ideas which we already hold, or the moral need to hold a different attitude. They may be wrong, but they are not vacuous. Typically, too, metaphysicians suggest that we ought to change, not just our strictly factual beliefs, but also our attitudes, the way we feel and act, both towards the world as a whole and towards particular things in it. Serious metaphysicians are reformers.

For example – Aristotle's thoughts on many particular matters converge to support the view that the intelligible form of the universe is divine, and he concludes that we ought to wonder at it and worship it. In this he was arguing against Plato, who had reasoned – again not just out of the air, but on clearly stated grounds – that human beings were essentially immaterial souls and should therefore try to live so as to escape from physical matter altogether. Kant, again, argued that human choice is free even though the world must be viewed as determined for the purposes of science, so that we should take our sense of individual responsibility much more seriously than we would if we genuinely thought of ourselves as cogs in a machine.

Similarly, Monod plainly expects his argument that the world is contingent not just to be intellectually accepted, but to change his readers' attitudes by freeing them from 'animism'. So did Hume. Hume, however, was a good deal more sophisticated and saw the dangers of competing openly in the metaphysical game when you also claim to be a sceptic. He said that he did not expect his arguments to change human life. Everything was to go on as before once the nonsense was exposed.[3] But he made it quite plain that he did intend to get rid of religion and the whole mass of follies that went with it. With the view he held of those follies, that change would actually have been drastic.

And so on. Very general metaphysical views like these are not just inert factual propositions, which we might accept without altering our attitudes or policies. They speak to our imaginations in a way that changes our world-pictures. They affect our symbolism. They reshape the framework of our thought. They shift our mental postures. They affect that whole vital central area of human life which connects thought, feeling and action. Though they are not themselves value-judgments, they do much to determine our value-judgments.

✳ TRUST

One very important way in which this works is through trust. In what do we put our trust? Do we trust the world? If so, what do we trust it to do? These questions may look odd, but they are quite substantial. Some people do trust the world more, some less. Though it may be tempting to say that we don't trust it at all, most of us have not fallen into the paralysis that that would involve.

The things we expect of the world vary considerably, and our expectations certainly affect our feelings and actions. The kind and degree of our trust varies, not just with our experience, but with our world-pictures. If we think of the world as a vast

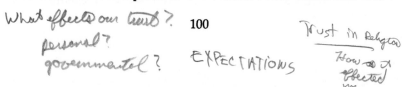

What effects our trust?
personal?
governmental?

100

EXPECTATIONS

Trust in Religion –
How is it effected me

machine, or as a contingent casino, or as an illusion that will some day vanish to reveal something else, this is certainly going to affect the kind and degree of the trust we place in it.

This kind of trust in the world can also be thought of as trust – or distrust – of our own powers. If we think of the world as ordered, we are perhaps partly expressing confidence in our own faculties as in tune with it and suited to find that order. Our faculties are, after all, part of the world and are not likely to reflect some quite different order from the rest of it. If, by contrast, we think of it as illusory, we may be thinking of our own faculties with despair as always hopelessly deceived. Or again, we may be trusting that we have certain faculties which can pierce through that screen, though others are misled by it.

The thought that we are looking inwards in this way – that we are dealing with trust in ourselves – may make the idea of trust seem less odd than it sounds if it is directed only outwards to the non-human world. I think there is something in this, though it is too simple an idea to explain the whole work of metaphysics. When we talk about the world as a whole, we are indeed necessarily also talking about our own place in it, about the relation we stand in to the rest of it, and about our own powers of dealing with that relation. These things all need to be considered together.

DARWIN'S SELECTIVE DOUBTS

This connexion between our trust in the world and our trust in our own powers is well lit up by a point about religious belief which Darwin raised in his *Autobiography*. Having dismissed fairly sharply the idea that religion must be true because many people believe it to be so, he goes on:

> Another source of conviction in the existence of God, connected with the reason and not with the feelings, impresses me as having much more weight. This follows from the *extreme difficulty or rather impossibility* of conceiving this immense and wonderful universe, including man with his capacity of looking far backwards and far into futurity, as the *result of blind chance or necessity*. When thus reflecting I feel compelled to look to a First Cause having an intelligent mind in some

degree analogous to that of man, and I deserve to be called a Theist.

Marginal note: This conclusion was strong in my mind about the time ... when I wrote the *Origin of Species*, and it is since that time that it has very gradually, and with many fluctuations, become weaker.

But then arises the doubt – can the mind of man, which has, I fully believe, been developed from a mind as low as that possessed by the lowest animal, be trusted when it draws such grand conclusions? May not these be the result of the connexion between cause and effect which strikes us as a necessary one, but probably depends merely on inherited experience? Nor must we overlook the probability of the constant inculcation of the belief in God on the minds of children having an effect that makes it as difficult for them to throw off a belief in God, as for a monkey to throw off its instinctive fear and hatred of a snake. I cannot pretend to throw the least light on such abstruse problems, and I for one must be content to remain an Agnostic.[4] (Emphases mine)

The interesting question is, why was Darwin's distrust so selective? Why was he suspicious only about the faculties which inclined him to believe, and not about the ones that inclined him not to? The reasons he gives for distrust are just the same in both cases. *All* our beliefs – including those that make us doubt God's existence, as well as those that support it – come to us through the faculties passed down to us through evolution. All have sources in our culture as well. And if one supposes that these two factors may be imperceptibly blurred through the inheritance of culturally acquired characteristics (as Darwin did), suspicion must become quite general. All our thought equally is then subject to undetectable corruption. – true

If we refuse this fatalistic conclusion – which, after all, is itself supported only by the corruptible process of our thought – we have to guard ourselves from error more selectively. We have to check one position by another and to look out for particular possible influences that may indeed have biased

then we would always be checking.

at what point do we stop + accept?

us. On the cultural scene, Darwin was of course right to note that belief in God had strong support. Psychological motives, however, often make us reject even views that are very strong in our culture. Darwin has just pointed out (p.50) that his father, his brother, and almost all his best friends were disbelievers. And he tells a story which is very illuminating about the meaning that this disbelief had for his father:

> Before I was engaged to be married, my father advised me to conceal carefully my doubts, for he said that he had known extreme misery thus caused with married persons ... My father added that he had known during his whole long life only three women who were sceptics; and it should be remembered that he knew well a multitude of persons and possessed extraordinary powers of winning confidence. When I asked him who the three women were, he had to own with respect to one of them, his sister-in-law, Kitty Wedgwood, that he had no evidence, only the vaguest hints, aided by the conviction that *so clear-sighted a woman could not be a believer.*[5]

doubt vs marriage

Gender issue alert!

MISOGYNY!

Women, in fact, are believers only because they are mostly so muddle-headed, but since that condition is incurable, they must usually be left to stew in their own juice. Charles Darwin himself does not endorse the distinct flavour of Enlightenment misogyny that rises from this and some other anecdotes about the old doctor. And he adds that, in his own time, sceptical women have become commoner. But he did, of course, have this same trouble with his wife, and he seems to accept fatalistically the psychological type-casting, the confused division of labour, that is the root of the trouble – women feel, men think. The intellect is a self-sufficient realm. Men never believe anything without reason, nor leave out any premises that women could show them. In a difference between the sexes, there cannot therefore be anything to be contributed intellectually from the women's position, the emotional position.

It is not seen that this is a conflict between two imaginative world-pictures, each of which has both an intellectual and an emotional aspect.

male dominated society

103

Emotional Intelligence - 'intellectualism' intertwined with emotion

Reason + emotion does one require the other

DOUBTS FROM EVOLUTIONARY THEORY

What follows, however, from the other worry that Darwin raises, from suspicion based on the lowly origin of our minds? Again, it is not clear how this attack can be made selective. *All* our faculties – not just our power of forming large positive speculations – have been 'developed from a mind as low as that possessed by the lowest animal'. Indeed, they have all been developed from no mind at all.

Do we need to get a certificate of how a faculty has been developed before we can use it? That would not be easy. Even on a confident view of evolutionary theory in general, we have only the vaguest idea of how our faculties can have become fitted to perform the ambitious tasks we set them. There is, for instance, a notorious puzzle about how our mathematical powers can possibly have been evolved. What conceivable use can they have been in the famous 'struggle for life'?

The same puzzle, however, arises about all our more complex powers of thought. For simpler ones, such as speech and elementary classification, we may think we do see clear evolutionary advantages. But in saying this, we are of course still using our more complex capacities. Without them, we could not consider the topic at all. And these complex powers are just the ones that raise Darwin's problem. His own theories about the workings of selection – which were relatively wide and undogmatic – provide only the most tentative cues towards explaining their presence. The much narrower, more 'sociobiological' proposals of his present-day successors are worse than useless on the matter, as Thomas Nagel has pointed out:

> The Darwinian theory of natural selection ... explains the selection among those organic possibilities that have been generated, but it does not explain the possibilities themselves ... (Even if we take those possibilities as given) ... In themselves, the advanced intellectual capacities of human beings, unlike many of their anatomical, perceptual and more basic cognitive features, are extremely poor candidates for evolutionary explanation, and would in fact be rendered highly suspect by such an explanation ... The capacity to form cosmological and subatomic theories takes us so far from the circumstances in which

our ability to think would have had to pass its evolutionary tests that there would be no reason whatever, stemming from the theory of evolution, to rely on it in extension from those subjects. *If, in fact, per impossibile, we came to believe that our capacity for objective theory were the product of natural selection, that would warrant serious skepticism about its results beyond a very limited and familiar range.* An evolutionary explanation of our theorizing faculty would provide absolutely no confirmation of its capacity to get at the truth.[6] (Emphasis mine)

Our ignorance about the source and biological function of our own powers of reasoning should not, then, make us suppose that we ought to give up using them. Nor did it stop Darwin writing his book. In general, this problem, as they say, *'solvitur ambulando'*. We prove that we can walk by walking. In general, Darwin too might accept this, but he is uneasy about using this casual, pragmatic approach when we come to 'such grand conclusions'. Perhaps we really do need a certificate of competence before we can touch on questions which are both very large and liable to stir reverence or wonder? In that case, as he sees, we must be just as cautious about denying those conclusions as we are about asserting them. Truth is equally beyond us either way.

HOW FAR CAN JUDGMENT BE SUSPENDED?

Darwin tried to remain neutral by being 'agnostic', by suspending judgment. That might have worked quite well if he had been dealing with a particular, detailed, unanswerable question such as whether certain distant unicorns exist. We would turn our attention to other things. But the question whether the sort of cosmic mind he means exists is not that kind of remote enquiry, cut off from the rest of life. It is a pervasive, important question about the whole world and our own relation to it, and indeed about our own nature. It is a major piece of metaphysics. It is therefore something on which we cannot, if we think at all, help having some kind of position. (We may indeed have two or more incompatible positions, but that is not the same thing as having none.) Refusing to pronounce about these things affects one's world-picture as much as pronouncing would.

Darwin surely knew this. It was what made him feel so deeply responsible for introducing a theory with vast and uncertain consequences. His tentative, reverent approach was entirely laudable; it is a mark of his greatness. It sets him right apart from some contemporary scientists who seem to have no doubt that, given enough research money, the largest imaginable questions can and will be settled by the methods of their own specialities.

This reverence, however, need not plunge us into the despairing scepticism he suggested. It does not mean that we cannot approach these questions, only that we must approach them in the right way. Items such as God and the soul are not unimaginably distant unicorns, about which we can have no evidence. There is evidence all round us, if we will use it carefully, in the world and in our own nature. The world as a whole is not something right outside our experience, unless we take a peculiar, narrow, solipsistic view of that experience as something outside the world. That kind of view was indeed suggested by Descartes's sharp division of mind from body, and still seems to many people to be a part of a scientific attitude. But it makes no sense and we had better avoid it.

(handwritten notes:) (i) the evidence for god inside us? (ii) that why many of us struggle w/ grasping this concept. We continually look to the material world for answers.

Reason ≠ acceptibility

Experience

10

THE UNINHABITABLE
VACUUM

ACCENTUATING THE NEGATIVE

Another much less impressive reason, besides Darwin's kind of laudable reverence, leads people to suspend judgment in this way. They believe that, on large subjects, it is always safer to be negative, to accept nothing that is not finally proved. Disbelief, as such, is then always preferable to belief, distrust to trust, scepticism to acceptance. Belief always shows weakness.

It is a main theme of this book that that idea is doomed because it is wildly and unconsciously selective. It always involves ignoring the mass of propositions we have chosen to believe before we start disbelieving. That selectivity was nicely shown in the title of a recent book *Can Scientists Believe?* which meant, of course, 'can they accept certain puzzling Christian doctrines?' The title quite overlooks the fact that difficult feats of believing more-or-less-incomprehensible things are an essential part of scientific work. *Can* scientists believe? Only watch them dealing with the more exotic parts of modern physics. As Bas von Frassen put it:

> once atoms had no color, now they also have no shape, place or volume ... There is a reason why metaphysics sounds so passé, so *vieux-jeu* today; for intellectually challenging perplexities and paradoxes it has been far surpassed by theoretical science. Do the concepts of the Trinity and the soul, haecceity, universals, prime matter, and potentiality baffle you? They pale beside the unimaginable otherness of closed space-time, event horizons, EPR correlations and bootstrap models.[1]

Should we meet this difficulty by conceding that indeed science is an exceptional case – an area where difficult belief really does become a duty – but adding that it is a duty because of a profound shift of intellectual values which withdraws that duty from all other areas? Everything not scientific should then preferably be disbelieved, though our natural weakness may stop us completing the task.

This is already a long way from the original indiscriminate position. But it is not far enough to provide a usable stopping-place. Science cannot stand alone. We cannot believe its propositions without first believing in a great many other startling things, such as the existence of the external world, the reliability of our senses, memory and informants, and the validity of logic. If we do believe in these things, we already have a world far wider than that of science.

The most crucial of those background things in which we need to believe is perhaps the conscious existence of other people. If we really doubted that – if we genuinely suspected that they might be a mere shadow-show, just empty behaviour-patterns or a set of robots programmed to delude us – then the testimony that they seem to give us about everything else would be worthless. The world would then shrink, not to the horizon of science, but to that of our own present field of consciousness. There could be no science without scientists.

Solipsism, meaning the belief that one is the only conscious being that exists, is a trap into which sceptical philosophers easily fall. It looks like a safe terminus for doubting, a haven where theorists can respectably sit and discuss at length the 'problem of other minds' – meaning the problem of how to prove that there are any after assuming that one is totally cut off from them. But solipsism is not really a terminus. There is no terminus. Even to be a solipsist you would need some beliefs; for instance, the belief that your self extends through time. You would also need some explanation of how language exists in an empty world. You would need, too, either to believe or disbelieve in those you argue with, and either course would have its drawbacks.

Spring-cleaning, in metaphysics as elsewhere, can become a confused obsession. C.S. Lewis has nicely traced its relentless progress from obvious usefulness towards final and hopeless incoherence. He writes:

At the outset the universe appears packed with will, intelligence, life and positive qualities; every tree is a nymph and every planet a god. Man himself is akin to the gods. The advance of knowledge gradually empties this rich and genial universe, first of its gods, then of its colours, smells, sounds and tastes, finally of solidity itself as solidity was originally imagined. As these items are taken from the world, they are transferred to the subjective side of the account; classified as our sensations, thoughts, images and emotions . . .

But the matter does not rest there. The same method which has emptied the world now proceeds to empty ourselves. The masters of the method soon announce that we were just as mistaken (and mistaken in much the same way) when we attributed 'souls' or 'selves' or 'minds' to human organisms as when we attributed dryads to the trees . . . Man is indeed akin to the gods, that is, he is no less phantasmal than they . . . There never was a Subjective account into which we could transfer the items that the Object had lost . . .

Now the trouble about this conclusion is not simply that it is unwelcome to our emotions. It is not unwelcome to them at all times or in all people . . . The real difficulty for most of us is more like a physical difficulty; we find it impossible to keep our minds, even for ten seconds at a stretch, twisted into the shape that this philosophy demands. And to do him justice, Hume (who is its great ancestor) warned us not to try. He recommended backgammon instead, and freely admitted that when, after a suitable dose, we returned to our theory, we should find it 'cold and strained and ridiculous'. And obviously, if we really must accept nihilism, that is how we shall have to live, just as, if we have diabetes, we must take insulin. But one would rather not have diabetes, and do without the insulin. If there should, after all, turn out to be any alternative to a philosophy that can be supported only by repeated (and presumably increasing) doses of backgammon, I suppose that most people would be glad to hear of it.[2]

SOCIAL COURAGE, TRUST AND DANGER

It is gradually being realized that this story of repeated evacuations is a mess. It has been built up by carefully ignoring, at each stage, prior and compensating movements the other way. How, asked Lévi-Strauss, could 'primitive man' ever have projected his own qualities 'anthropomorphically' on to natural forces,

> without simultaneously making the opposite move of attributing a power and efficacy comparable to that of natural phenomena to his own actions? This man, externalised by man, can serve to shape a god only if the forces of nature have already been internalised in him. The mistake made by Comte and the majority of his successors was to believe that man could at all plausibly have peopled nature with wills comparable to his own without ascribing some of the attributes of this nature, in which he detected himself, to his desires.[3]

Naive perception, in fact, is in the first place of the outside world; without that, no notions of the self would ever have been formed to be projected.[4] The spring-cleaning campaign is fraudulent from the start. As Lewis points out, the stage at which it most obviously loses touch with reality is in trying to scrub away our convictions about the solidity and complexity of other people, and our confidence that it is possible to know them.

It is worth noticing here how oddly the disapproving attitude to belief which we have just mentioned relates to our view of trust. Though it is possible to be too trusting, someone who systematically distrusts people rather than trusting them does not strike us as an admirable or sensible character. Some degree of social courage – the willingness to risk being hurt in order to get near to people, to risk being misled in order to communicate – is an essential cognitive tool. It is also a necessary virtue, since the things that need doing for people cannot be done if you are too scared to go near them.

There is no real safety in being negative. As William James pointed out, a negative choice commits the chooser just as much as a positive one does. He put the point in a discussion about acceptance of religion, but what he says applies to all propositions of wide importance:

> We cannot escape the issue by remaining sceptical and

waiting for more light, because, although we do avoid error in that way *if religion be untrue*, we lose the good, *if it be true*, just as certainly as if we positively chose to disbelieve. It is as if a man should hesitate indefinitely to ask a certain woman to marry him because he was not perfectly sure that she would prove an angel after he brought her home.

(Here again, surely, the idea of a sex-linked relation between feeling and reason is active?)

To preach scepticism as a duty to us until 'sufficient evidence' for religion be found, is tantamount therefore to telling us . . . that to yield to our fear of being in error is wiser and better than to yield to our hope that it may be true. *It is not intellect against all passion; it is only intellect with one passion laying down the law* . . . What proof is there that dupery through hope is so much worse than dupery through fear?[5] (Last emphasis mine)

As James points out, a chronically timid, suspicious temper may have its uses, but it is certainly not an infallible guide. There is, moreover, at least one important range of situations – namely, the social one – where it can actually prevent our ever discovering truths at all:

The universe is no longer a mere *It* to us, but a *Thou*, if we are religious; and any relation that may be possible from person to person might be possible here . . . To take a trivial illustration; just as a man who in a company of gentlemen made no advances, asked for a warrant for every concession, and believed no one's word without proof, would cut himself off by such churlishness from all the social rewards that a more trusting spirit would earn, so here, one who should shut himself up in snarling logicality and try to make the gods extort his recognition willy-nilly, or not get it at all, might cut himself off for ever from his only chance of making the gods' acquaintance.[6]

There is, then, a vast range of social facts – a range containing perhaps all the really important social facts – which this radically distrustful person would never discover. Are we sure that the same thing is not true of many facts that are not obviously

111

social, and especially of facts about the universe at large? May the discovery of these too not require a certain kind of trust?

UNDERSTANDING, TRUST AND RESPECT

It will not do to dodge that question by flatly replying that the universe is actually just a non-social *It*, that James is simply mistaken in treating it as a *Thou*. As we have seen in the last few chapters, the category of *It*, or inert, brute matter, is in our tradition a thoroughly confused one. The category of *Thou* has undergone corresponding confusion, and attempts have been made to confine it very narrowly to the human intellect and will. But it surely needs to be much wider, and should be understood as having much more variation within it, many more kinds and degrees of application.

The sharp, binary division into *persons* and *things* did excellent work when Kant first introduced it, because he used it solely on its positive side, to emphasize the special respect owed to rational human beings. Persons, he said, must never be treated solely as means, but always as ends in themselves also. Because they have reason, they have dignity; they should be free to guide their lives by their own thoughts.

On the human scene, this proved a very effective argument, not only against oppressive institutions such as slavery, but also for civic freedom against benevolent despotism. Difficulties arise, however, when we turn to the other side of the coin and ask, 'Can all non-persons, then, be exploited without limit? Have they no dignity, are they never entitled to any respect?'

Kant himself had trouble explaining why cruel treatment of animals was wrong (though he thought that it was).[7] Since his time the artificial limitation of our official value-system to the human scene – indeed often to the political scene – has made increasingly grave trouble, as we are now finding when we look for words to explain why we ought to respect the biosphere. It may well seem better to talk at least, as Albert Schweitzer did, about 'reverence for life'. But is it possible to stop there? As Kant himself said – disregarding Descartes – 'Two things fill the mind with ever-increasing wonder and awe the more intensely the mind is drawn to them; the starry heavens above and the moral law within'.[8] Is there anything wrong with that? And does not our natural sense of wonder legitimately extend to many things on earth, as well as in the heavens?

We can easily see that it does without getting on to the debatable ground of religion. We only need to think of the respect and sympathy with which a true worker in any craft treats tools and materials. Engineers and joiners do not see wood, stone, earth, water and metal just as passive, alien stuff to be kicked around. They understand that they are interacting with it. They study it, they 'have a feeling' for it – just that feeling which the rest of us show we lack when we helplessly bang our apparently inert alarm-clocks.

This is not only true of crafts. It is still more obviously true of those who work with living plants and animals. A remarkable example has been the advances made recently in our understanding of primate behaviour by enquirers like Jane Goodall, who have shown that a sympathetic, respectful approach to the animals is far more productive, even in terms of simple information, than the traditional more remote and defensive one. By trusting their animals – by assuming that a relationship is possible – these people earn trust in return, and are able to enter into that relationship in a way that makes possible much more sophisticated communication.

Of course this policy of trust does have its dangers. Trusting people do sometimes get bitten. The chance of getting bitten is indeed the price that must be paid for getting near to any other organism. One can get dangerously bitten by other people too, if one is rash enough to trust them. But people who will not take that sort of risk can never hope to achieve very much in the way of understanding, let alone of relationship.

In theoretical enquiry, too, the same difficulty crops up over all attempts to understand anything unfamiliar. To treat our subject-matter as mere brute, alien object always hinders the work. Scholars who insist on working on topics that they do not really like commonly work badly. Though we must of course avoid superstitious attempts to over-identify with our subject-matter, it is no use trying to keep it always at a distance. Our imagination vitally needs in some way to touch it, to leap inside it, to try and see how things work from that alien point of view. That imagination is the only tool we have for starting the job. Certainly it has faults, but the way to correct them is by using it more fully and carefully, so as to make it correct itself. If, instead, we refuse to use it at all because we distrust it on principle, we might as well give up and shoot ourselves right away.

PRAGMATISM GOES HALF-WAY

This is not a simple point. It may be helpful to illustrate it by looking at a half-way position. William James, in attempting to correct a proposal of neutrality like Darwin's, began splendidly, but went wrong, I think, by not going far enough. He still kept the traditional division of the mind into two warring elements, Feeling and Reason. Like Hume, he tried to put the imbalance right by simply giving the victory to Feeling:

> Our passional nature not only lawfully may, but must, decide an option between propositions, whenever it is a genuine option that cannot by its nature be decided on intellectual grounds; for to say, under such circumstances, 'Do not decide but leave the question open' is itself a passional decision, – just like deciding yes or no – and is attended with the same risk of losing the truth.[9]

This treats the issue as though there were two separate officials, Intellect and Passion (easily seen as male and female) who cannot work together, but must share out questions between them and deal with them separately. This would be an odd arrangement even for two separate people; for the two aspects of one person it is extraordinary. The idea is, of course, that questions come in two distinct kinds, intellectual and passional, needing quite separate processes to resolve them. This is quite unconvincing. It is a fact of the first importance about questions that – if they are difficult at all – they are nearly always complex. They have a number of distinct aspects needing different kinds of thought and attention. They have to be solved by co-operation. They need an integrated personality.

William James's method led him to take sides ardently with the passional element as the more fundamental. He noted many genuine and interesting ways in which particular motives do indeed quite legitimately influence our thinking. In this way he showed how to correct the gross leaning which our tradition has shown towards the opposite error, and this is indeed very helpful for integration. But he himself remained no less one-sided than his intellectualist opponents. He flatly defined truth in terms of usefulness, and subordinated theoretical curiosity to the basic choice of aims.

Though there is indeed a vital half-truth here, it was one that the times could not receive. Accordingly, in an age that almost deifies theoretical curiosity, his arguments have been side-tracked. His position tends to be dismissed as morally feeble, as an indefinite licence to invent unicorns, as an escape into wish-fulfilment from the harsh challenge of facts. Its force and subtlety get missed. I shall repeatedly be reworking his points in an attempt to deal with this difficulty.

11

PARSIMONY, INTEGRITY AND PURITANISM

WHAT ARE WE SCEPTICAL ABOUT?

Many people in Darwin's day naturally shared his dilemma about how to see the world as a whole realistically. As time went on, they increasingly felt that the supernatural element might be an unnecessary extra, a frill added to the natural world, a wish-fulfilment springing from human childishness and self-indulgence. The supernatural seemed, as Stephen Jay Gould lately put it, 'the representation of raw hope gussied up as rationalized reality'.[1] The moral issue was then clear. Religion – meaning Christianity – was an addiction that honourable people must break. And if that left the world without meaning, then meaning must be somehow rebuilt or finally dispensed with.

In attempting this rebuilding, sages made much use of science. The beauty and order of the Newtonian universe seemed to offer a new home to exiles from the traditional Eden. It gradually became clear, however, that this beauty and order might not be any more secure than what they had replaced. Only in patches is the order and beauty of the world directly visible. To believe in it as a whole requires faith.

Up till now that faith – that conviction of a universal order – had been backed by and expressed in belief in God. Without that, what remains is just the conviction of scientists that the world must finally conform to science – that doubts and confusions will eventually give way, revealing underlying order. But might not that faith too be mere wish-fulfilment? Is there any real guarantee that the partial order we have seen so far is not a misleading varnish on hidden disorder? Or that, even if it has been real so far, it will not stop next week? Or that it covers matters we have not yet experienced? Induction makes us expect continuance, but what justifies induction? As Russell pointed out, 'The man who has fed the chicken every day throughout its life at last wrings its neck instead, showing that more refined views as to the uniformity of nature would have been useful to the chicken.'[2]

PROBLEMS OF PURE THINKING

Darwin's story illustrates the very selective way in which what we may call heroic Enlightenment puritanism works. Not all wishes get mortified, only certain chosen ones. Puritanism – when it is used as a term of criticism or abuse – means objecting to pleasant or easy things merely because they are pleasant or easy. This can lead to counting things as temptations which ought not to be so counted.

In the case of thought, puritanism calls on us sternly to avoid wish-fulfilment – to abstain from accepting any belief merely because we would prefer it to be true. But, as many people have pointed out, we do prefer to believe the world to be intelligible. Have we a right to indulge that preference?

IS THERE ANYONE UP THERE?

There are two ideas here which we might like to find true – first, that the world is in fact ordered, and second, that that order is accessible to us, which is naturally taken to mean that it is an expression of a mind like our own, but much greater. The first idea has its own problems, as we shall soon find. But it is the second that really needs attention. Its critics often assume, as Gould does, that this is obviously just a piece of wish-fulfilment, a couch for the lazy-minded. In more hierarchical ages, that may well have been true. But moral and political taste and imagery have changed profoundly in the last three centuries, and it is no longer true today.

In our current individualistic climate, the idea of having an authority above us who always knows best is far from welcome. The virtues we are taught to revere most are no longer ones appropriate to subordinates, such as patience, loyalty and obedience, but ones fit for solitaries or rulers, like autonomy, independence and moral courage. (Of course these are officially now meant to fit us for life as equals, but that is something of a pious hope.) We find it hard enough to relate to directors set above us here on earth; it is far harder to know what to do with a heavenly superior, even a very remote and abstract one. That change of moral temper, and not any scientific discovery, seems to me the root cause of the modern estrangement from traditional religion.

Changes of moral tone like this are, of course, never complete; they are matters of emphasis, which is just as well. Both for political and private purposes, both these attitudes are always needed, and are always present. Complete individualism would be as unworkable as complete conformity. Yet changes of emphasis do have a marked effect, and the upvaluing of individual pride is evidently making trouble for us in accepting a subordinate cosmic status that did not trouble our less assertive ancestors at all. They simply did not feel the need to think of themselves as ranking first in the universe.

This is surely the change that has called for a general massacre of the supernatural, a purge of the wide and varied fauna that normally inhabits the human imagination. The root cause is not any shift in the nature of the explicit arguments. These remain as vast, difficult and confusing as ever, much hampered by the lack of an adequate language to express them. They are not what really determines people's faith. Neither, however, is the cause of change a moral reform of the kind usually suggested, namely a sterner mood of resolution among scientists and other sages, enabling them to cast off chains that earlier generations dared not shift.

The position is not that theists such as Newton and Galileo were too ignorant and childish to see what was obvious to T.H. Huxley and Bertrand Russell, nor that they were too dishonest to admit it. Neither is it that new facts, discovered since their day, have put their theistic attitudes out of date. (They would not have been among the people who were surprised when the first Soviet astronauts failed to find God in outer space.) It is true, of course, that their views were influenced by the moral and political climate of their times. *But then, so are ours.*

Contemporary climate always makes a great difference to what counts as a temptation. It may indeed be characteristic of small children to want a familiar system and to depend on having someone above them. It is also characteristic of adolescents to throw off that system and to protest against those guardians. Moving from the first state to the second may indeed be progress, but it is not yet maturity. Mature people are supposed to be relaxed about questions of rank and to look for such system as suits the general good, whether familiar or otherwise. A passionate conviction that there cannot possibly be a mind superior to one's own is certainly not something that can be drawn from science. It seems to me much more like what Gould briskly calls the representation of raw hope, gussied up as rationalized reality.

ARE WE ENTITLED TO ORDER?

If, however, some hope is raw, might there be other kinds of hope that are cooked, processed, somehow tested and found reasonable? We do, as I pointed out, naturally hope that the world is orderly. We like it that way. As we have seen, this idea of a basically ordered world is even one which, today, may be very important to us emotionally, may seem an important aspect of our salvation. All of us, including those ignorant of science, find this idea sustaining. It controls confusion, it makes the world seem more intelligible. But suppose the world should happen in fact to be *not* very intelligible? Or suppose merely that we do not know it to be so? Might it not then be our duty to admit these distressing facts?

This is a real difficulty. We are all children of the Enlightenment, whatever other forebears we may acknowledge. It has been a cardinal principle of our upbringing that we must never believe things simply because we want them to be true. But how are we to apply that principle to cases where our wanting-them-to-be-true is essentially a matter of the satisfaction of reason?

Rationalist thinkers – ones who trust our natural leaning towards an intelligible world – have always argued that the real world must indeed be perfectly rational. This assumption, however, has gradually become discredited because the ideas of various rationalists about what actually was rational conflicted and led to rather irrational disputing. Kant, for instance, thought that Euclidean geometry was a necessary form dictated by the nature of rationality, a shape to which reality must conform. Hegel, notoriously, proved on rational principles that there could only be seven planets . . . And so forth.

This simple, confident rationalism, in fact, is officially dead. But that does not stop much of science – especially some kinds of science, especially theoretical physics – from going on very much as if it were true. When cosmological theory advances, and rules that something must have been the case about the Big Bang, it is generally assumed that this is not just a move in a highbrow game played in physics labs, but the discovery of a fact about the real world. It is not an empirically discovered fact, and there need not be any direct physical evidence for it. It rests on the logical coherence of theory.

If we ask a physicist why this reliance on human reason is all right, when Kant's and Hegel's and indeed Marx's reliance on their reason led them so wrong, he is likely to say that there is nothing objectionable about relying on reason as such; these people just reasoned wrongly. We are then (more generally) justified in assuming regularity in nature. Main current theories can be assumed to be true directly of the world.

This cheerful assumption of harmony between our faculties and reality – known as 'naive realism' – is the attitude which scientists normally hold, and perhaps must hold, about the work they are doing while they are doing it, just as the rest of us do about whatever we are dealing with. Most of the time this faith is perfectly satisfactory, but occasionally it begins to matter how ambitious the claims involved are.

At the modest extreme, our 'realism' might only mean that we believe the thing we are talking about is really there. At the more ambitious pole, it might mean that we believe it is exactly as we describe it. This ambitious claim gets less and less convincing as one conceptual scheme after another has to be changed. Our ancestors were, we think, seriously mistaken in classing whales as fish or oxygen as dephlogisticated air. But of course they were not wholly wrong; they had grasped some part of the truth. The whales were indeed not birds nor the oxygen a form of brickdust.

Radically sceptical suggestions that all our knowledge is just a social construction, not shaped at all by anything outside us, do not make much sense. We need a workable compromise. We need somehow to insist that the world really is there, and that we are not wholly mistaken about its nature, while admitting that our acquaintance with it is slight and patchy, blurred by all sorts of cognitive weaknesses. 'Realism' is not the name of one among two rival alternatives. It is not a football team for which one can sign up. It is one pole of two between which we must somehow find a place.

How far, then, can scientists be realists? In some sense, as just mentioned, they must be naive realists when they are actually working. When, however, they feel sceptical doubts, they can put on a different hat and use a more cautious attitude – some chosen form of positivism or instrumentalism. They can say that what they seem to be saying about the world is not meant literally, but is only an explanatory construction summing up past observations so as to predict future ones. As Rom Harré puts it:

> Instrumentalism . . . advocates the view that theories are not to come up for judgment as true or false, indeed, they cannot so come up, but are to be judged by whether they are successful or unsuccessful 'instruments' for research.[3]

This is, as he points out, quite an old view. Copernicus's posthumous book *De Revolutionibus* contains a preface, added by its nervous editor, to explain that its alarming doctrines need not be taken as literal truth. Similarly, Bishop Berkeley wrote that asking whether the earth moves is really only asking

> whether we have reason to conclude, from what has been observed by astronomers, that, *if* we were placed in such and such circumstances, and such and such a position and distance both from the earth and sun, *we should perceive* the former to move among the choir of the planets, and appearing in all respects like one of them.[4] (Emphasis mine)

Scientists need not, then, believe that anything outside the minds of observers exists at all. When they seem to be speaking about the movement of remote stars, or about the behaviour of imperceptible particles, they are not really referring to those external things. They are just describing the hypothetical experiences that human observers would have in certain circumstances. 'Particles', 'galaxies' and the like are only the names of concepts convenient to operate with, not the names of real objects at all.

PURELY SPIRITUAL EXISTENCE

This kind of view posed no great problem for Berkeley, who held that all reality was spiritual anyway. For him, the unobserved outside world really did not exist except as a set of ideas in the mind of God. Berkeley was an 'idealist', in the full metaphysical sense that involves (not high moral ideals, but) the rejection of matter. He had paid his metaphysical entrance-money for his remarkable view about science. J.-L. Borges, who is fascinated by Berkeley, has written a story about a world of this kind, where people really do not believe in objects which are not being perceived:

> The world for them is not a concourse of objects in space; it is a heterogeneous series of independent acts . . . There are no nouns . . . There are impersonal verbs, modified by monosyllabic prefixes . . . For example, there is no word corresponding to the word 'moon', but there is a verb which in English would be 'to moon' or 'to moonate'. 'The moon rose over the river' is *hlör u fang axaxaxas mlö*, or literally: 'upward behind the far-streaming it mooned'.[5]

For Borges's people, the idea that unobserved objects might continue to exist is an absurd, inconceivable paradox . . .

There is plenty of point in such a story, but this state of things surely demands unquestioning belief in a spiritual world of a peculiarly solid sort. This is not just a distinct culture; it is (as Borges says) a different world, with a kind of imagination that may indeed not be human at all. There would have to be a strong supernatural background to supply the continuity that is lost from the world if moons, tables and human bodies really only exist at times when they are observed.

Berkeley's ideas, however, have been taken up by theorists who did not always notice what drastic changes they called for. Nineteenth-century thinkers, following Ernst Mach, adopted this notion as a general explanation of scientific claims about the external world, often without paying the extra and accepting metaphysical idealism. Instrumentalism was transplanted without the soil in which it had originally grown. Their successors, less and less interested in metaphysics, have become even less aware that they were doing this.

Without God as a receptacle for the vast unobserved background, Berkeley's suggestion looks much odder. If it is then to become something which can actually be believed, rather than just being recited in order to get out of a difficulty, it needs some different kind of metaphysical support, which no one seems yet to have supplied. Since, however, these larger implications are often not noticed, instrumentalism is seen as metaphysically economical, as a modest, parsimonious way of avoiding saying anything about unobserved entities. It has been an important source of the modest, minimalist image of science which we noticed in chapter 1.

Like other modest, minimalist claims, instrumentalism looks good when you want to keep out of trouble, but it is impossible to live up to it in normal life. As just mentioned, most scientists simply forget it most of the time. They assume that they are talking about the real world directly. Without this belief, they, like the rest of us, would be liable to find their imaginations totally disoriented, so they quietly use it. For the more reflective among them, however, this double life is uncomfortable.

That is probably why some of them have lately turned to another way out, saying that the universe itself is in some strange sense the product of our minds – that we produce it by observing it, that it is simply a mass of information, validated by and continuous with the mass that we have in our own thoughts. (Information, we may note, is the modern substitute for spirit.) Thus a curious, ill-understood kind of idealism has lately been creeping back, suggesting that the universe is in some bizarre way our creation. No doubt this is calculated to make it seem less surprising that it should conform to the laws of our thought. We will be looking further at this strange neo-Berkeleyan view later.[6]

THE PROBLEM OF AIMS

Altogether, then, the difficulty about justifying scientific confidence is a real one. If this confidence is all right, what makes it so? As I have tried to show, this is not just a question about the nature of the physical world, nor about the possibility of knowing that world. These questions about knowledge cannot be properly approached without raising central questions about what James called our 'passional nature', about what matters to us, about the general aim of our lives.

Knowledge is not an isolated phenomenon. It is made possible by trust, and we do have a choice about what we will trust. The immense achievements of modern science have grown directly from a revolutionary decision to trust the physical world – to assume that it had an underlying order. This involved a corresponding decision to trust those faculties in ourselves that serve to seek such an order. At the Enlightenment, we – Western intellectuals – took a resolution to trust those faculties above all, rather than other faculties which had previously seemed people's surest guides, faculties centring on their moral being and their own self-knowledge. That, I think, is what is centrally meant by calling the age that has followed a scientific one.

What, however, are the faculties in which such a scientific age puts its trust? They are often described as being sceptical, critical and methodical. They are taken to centre on negation – on discipline, on the control of thought, on resisting natural errors. But this overlooks the immense, primary, positive act of trust directed to the physical universe that was necessary before these critical skills could be used. Highly disciplined, sceptical thinking was not itself a novelty, certainly not something peculiar to the modern age. There were very sophisticated sceptics in ancient Greece, and also among late medieval philosophers. That scepticism, indeed, was what built up the impression of futile cleverness which finally discredited scholasticism. Descartes's distinctive move, which made room for Galileo and for modern physics, was not his scepticism, but his finding a way out of that scepticism by a radical act of trust in scientific reason.

That act of trust was explicitly a religious one. Descartes argued that we can trust the Creation because we can trust God the Creator. He defended that position by complex arguments, carefully related to the methods of science. But arguments of this kind turned out ill-suited to this work. The kind of God they described was inevitably something more like a natural force than like the God that people experienced in their lives. As time went on, this 'God of the philosophers' became more and more of a distant abstraction, and it became less clear why there was any need to add him to other scientific concepts. Thus, finally, Laplace told Napoleon that 'he had no need of this hypothesis'.

As far as the internal success of science was concerned, this was of course understandable. At that time, Newtonian and mechanistic thinking looked complete and self-justifying. But that fact is no help towards the question 'what, fundamentally do you put your trust in?' Such trust is not just a matter of procuring information. It is a matter of profound reliance, of what one believes lies under the surface of life, what will endure when that is shaken.

If all religious and contemplative regard for the world and its creator are removed from the scientific attitude, we may be left answering simply, 'Negation remains. You can always trust your critical faculties.' The trouble with this is not just that it is depressing, but that it is false. Critical faculties, left to themselves with nothing positive to criticize, cannot function at all. They have no standard to work by. They rapidly eat themselves away and disappear up their own orifices, as they have always done in purely sceptical thought.

Moreover, the idea that we should rely solely on them – the sense that somehow we *ought* to – is itself a moral judgment, and a very remarkable one, which those critical faculties themselves cannot validate. Taking this line would not release us from the need to treat this whole issue as what it is – a moral one, a question about how we should aim to live. This is what is still not sufficiently noticed about the change to a 'scientific age'. It has been above all a *moral* change. A change in what we trust has inevitably involved a change in what we admire and honour, therefore in the direction we give to our lives.

12

QUESTIONS OF MOTIVATION

OUGHT WE TO KNOW ANYTHING?

Psychological factors (handwritten margin note)

What does this change mean psychologically? Throughout this book, I have been directing attention to psychological factors – to motivation, to comprehensive world-pictures, to the myths, dramas and fantasies that form the imaginative background of all our intellectual work. It is in this context that we need to ask: What qualities of character, what attitudes to one's own feelings, are really required for a scientific approach? The asceticism of the Enlightenment vision, and the ideal of parsimony that is supposed to inform it, must be our next topic.

Two motives which are both thought of as praiseworthy and necessary, not only for science but for all enquiry, seem to conflict here. Our desire for understanding appears to clash with the need for courage in admitting disagreeable facts. As Hume and other sceptics have pointed out, it is not simply obvious what justification the comfortable, welcome idea of the absolute regularity of nature has – apart from the fact that we like it.

What I want to draw attention to here is not the problem of justification itself – the problem of how knowledge is possible – but the question of how it should be regarded, the drama that has always been felt to attach to it, and in particular the moral pretensions of the disputants. I think we should be interested in asking why, when we enquire about the nature of knowledge, a suspicious, disbelieving temper has been held to be, simply in itself, so much more morally respectable than a hospitable, believing one.

The more technical, limited problem about why we should

accept the regularity of nature has, of course, been the subject of much philosophical dispute. I am happy to notice that the general upshot today is that the terms of the question were, in one way or another, mistaken in the first place. This is surely right. There has been far too much very general epistemology done because there has been a bad habit of dealing with such questions in wildly abstract terms, a habit of false universality. The very sweeping attempts to attack, or to 'justify', induction or other aspects of knowledge as a whole, which have absorbed so much time and ink since Descartes's day, have been misguided. The meaning of claims to knowledge varies hugely from one situation to another, and the kinds of attacks and defences that make sense about them have to be calculated for those different situations.[1]

The regularity of nature was of course a natural focus for these disputes. Rationalists have suggested all sorts of arguments to prove that regularity had to be viewed as necessary against the background of a wider conceptual scheme. But – as Hume pointed out – it is always possible to ask whether we need accept this conceptual scheme as a whole. If we do ask this, one reason why we shall refuse to abandon it undoubtedly is an emotional one, and it is a reason so strong that it alone would make the thing impossible.

To live without any belief in a fixed order around us would be horrifying. We probably couldn't live that way at all, and we want to go on living. Of course it might be possible to live with a belief in a rather looser order, an order inadequate for science. Many peoples probably do live with that kind of slighter belief. But for us, to live that way would mean doing without science. And science is something that has been built into our culture, yielding us real aesthetic and emotional benefits. If a more drastic, thorough destruction of belief were demanded, it is not clear that our brains are so constructed as to allow it. Apart from the alarm produced by a belief that causal connexions all round us were never really reliable, it would cost us a terrific, constant effort to keep reminding ourselves to treat them in this way. Hume said that we are simply too lazy to attempt this:

If we believe that fire warms or water refreshes, it is only because it costs us too much pains to think otherwise.[2]

PURITANICAL BLACKMAIL

This seems much like saying that it costs us too much pains to walk on our hands rather than on our feet. It does, and that is a perfectly good reason for not doing it. Notice Hume's insidious dramatization, the aggressive moral tone. Scholars do not like being called lazy. Hume's arguments have been answered often enough, but the moralistic drama that surrounds them has not had half enough attention. He managed to convey that people ought to feel guilty if they believe anything more than the minimum, that anyone believing anything is always one down morally on people who disbelieve it.

This sense of guilt still persists, even though Hume did not actually say that we should stop believing in the physical world and its regularity. He told us to go on doing so as before, but not to suppose that we had any good reason for it:

My intention then in displaying so carefully the arguments of that fantastic sect [sceptics] is only to make the reader sensible of the truth of my hypothesis, *that all our reasonings concerning cause and effect are derived from nothing but custom, and that belief is more properly an act of the sensitive than of the cogitative part of our natures.*[3] (Emphasis Hume's)

The trouble with this is that he is not neutral about it, nor is he – at heart, as they say – really a partisan of the Sensitive Part. He still speaks as a dyed-in-the-wool representative of the 'age of reason'. His argumentative tone is so strong, his contempt for undefended positions so chronic, that belief still comes out sounding like a weakness. As with William James later, the sharp division of the human personality into a cogitative and a sensitive side makes it impossible to do proper justice to the sensitive one, even for writers like these who are actively trying to.

This feeling that disbelief always has the right of way still persists; it can be noticed in almost any philosophical controversy. Of course it was originally a natural reaction against the reverse situation. Hume lived all his life under the shadow of the dogmatic Calvinist Kirk, in a town where it was dangerous to admit that one disbelieved certain Christian doctrines. He was quite right to fight back. And it is not surprising that he fought back with his opponents' weapon: dogmatic self-righteousness.

Background conditions, however, change. Today, it is the admission, not the denial, of belief in central Christian doctrines that can damage the reputation of an academic in Britain. And more generally, it is the admission, not the ignoring, of the part played by feeling in thought that still alarms the academic mind.

We need to say firmly and repeatedly, against Hume, and also against the tide of our times, that the mere presence of an emotional factor in any kind of decision does not take it out of the realm of thought. All our thinking involves emotional factors as well as rational ones, just as every physical object has size as well as shape. These are not alternatives. The presence of one doesn't mean the absence of the other. The kind of emotional need that we have to see the universe as ordered is not something alien to thought, nor is it only its biological cause. It is also its conceptual condition. The need is a single need with two aspects.

More deeply, this whole cleft between feeling and reason – this official division of our nature into radically distinct emotional and rational elements – with which European philosophy long worked and which Hume sharpened to the point of suicide, is a disastrous error. It hides essential organic connexions in the middle ground, structures common to our thoughts and feelings. And this middle ground is specially important for very large metaphysical questions concerning things like the kind of order that we need to believe in.[4]

Plainly, it is not clear how far this general suggestion can take us. At present I am making the point at a fairly simple level about the mere minimum belief in an order sufficient to secure the facts of science, something which I guess most of us would like to see secured, and a matter on which – in spite of Hume's blackmail – we don't regard our wishes as immoral. I am inclined myself to think that the general point must take us a good deal further than this. To believe in order at all is to have a belief about the nature of the whole that goes beyond simply securing those individual facts.

IS IT ALL ALL RIGHT REALLY?

To come back, however, for the moment to our analysis of heroic Enlightenment puritanism – this worry about possible illicit wish-fulfilment in our acceptance of scientific order is not new. It has disturbed the conscience of intellectual puritans ever since Hume's time, and indeed before. (Many of the sceptical arguments are much older.) Hume, when he made remarks like this, added at times – as sceptics tend to do – that people ought not to be disturbed by his attacks, because everything could go on just as before. Belief was indeed being reclassified as merely a kind of feeling, but this need not upset our confidence in its validity. He was only saying that 'belief is more properly an act of the sensitive, rather than the cogitative, part of our natures'.

These attempts to draw the sting have, however, not managed to reassure his readers. One reason for this is of course Hume's style – the cocky, triumphant reductivism of remarks like the one just quoted and plenty more – for instance and typically:

> It is impossible, upon any system, to defend either our understanding or our senses; and we but expose them further when we endeavour to justify them in that manner. As the sceptical doubt arises naturally from a profound and intense reflection on those subjects, it always increases the further we carry our reflections ... Carelessness and inattention alone can afford us any remedy.[5]

Nobody who did not mean to change people's practical attitudes, to have a serious subversive effect, would write in that dramatic style. But there is a substantial reason as well why it is not possible to accept Hume's reassuring remarks. *There simply is no way in which we could restate all our beliefs in the language of feeling* – could reclassify them in the way feelings are classified and judge them by the standards appropriate for feelings. Any serious attempt to do this kind of thing results in distorting the language of feeling itself – as happened indeed when Hume himself tried to do a little of it for morals by rephrasing moral reasoning in terms of the Sentiment of Humanity.[6]

DEFENCES – POSITIVISM AND PRAGMATISM

Altogether, then, the charge that claims for the truth of science have been inflated by irrational wish-fulfilment has been widely held to be a serious one, and people have wanted to answer it. For this purpose, apologetic self-denying ordinances have been devised which are meant to make these claims to truth look more modest. First, there has been the Positivist or Operationalist suggestion already mentioned, that the findings of science are not really meant to be literally true at all, but are just convenient ways of summarizing the data. Thus, if I say that the continents are moving, or that birds are descended from dinosaurs, I can't really mean these things, because they are matters too large for me to know. I am simply using these phrases as handy formulae for summing up the detailed observations that suggest those possibilities.

This might be all right if it didn't have to be done again for each of the separate observations in turn, leaving us eventually plunged in Phenomenalism, explaining that we are really talking about nothing but the subjective experiences of the various observers. Since these experiences are usually quite unknown to the people who use the large formulae, this is not very convincing.

Or again, worried people have taken refuge in Jamesian Pragmatism, saying that the claim to truth is harmless because truth simply means usefulness. In that case, if I say it is *true* that the continents move, or that birds are descended from dinosaurs, I am really only saying that it is useful to go on as if these things were so. This seems much like saying that, if it is useful for people to believe that there is a ferocious demon inside the electric wires – a belief that stops them touching the wires and getting electrocuted – then that usefulness is the same thing as its being true. This is not very plausible. And anyway, how do I know whether it is true that these beliefs are even useful?

Both these lines can be made much more subtle to meet these difficulties, and some highly intelligent people have devoted their lives to making them so. Even when this has been done, however, these devices are not going to meet the present danger unless they compel us to *mean* something quite different when we make these statements – not just when we make scientific statements, but every time that we make ordinary everyday statements as well. It looks as if there will have to be some kind of vast, chronic withdrawal of confidence, a withdrawal to a position which is itself obscure and hard to find.

This is not just a difficult move, it is an unreasonable one. Of course it is true to say that we often ought to be less confident in our judgments than we are. Indeed, it wouldn't hurt some of us to be less confident always. But that is because it has often been shown that these judgments are *wrong*. For this to happen, we have to have confidence in the evidence that refutes them, and in the reasons that support that evidence. Confidence as such hasn't been abolished or even weakened, it has just been shifted. This shifting of confidence among the various things that we believe is common, and can have most important consequences. But a universal, indiscriminate weakening of confidence doesn't seem to have any such consequences or meaning at all.

WHY PURITANISM?

On the whole, then, there seems no reason to join in these rather furtive and guilty defences that have been devised against the attacks of heroic intellectual puritanism. It is surely better to stop being apologetic, to go on the offensive and to ask heroic puritanism, in turn, what are its own credentials.

Puritanism simply as such is, after all, not something that can count on the respect of the present age. In general, people today do not accept that the mere fact of their wanting something is itself an objection to their having it. Is there an exception to this for beliefs? Is there really a general reason why we should believe as little as possible, though there is no parallel reason why we should eat as little as possible or take as little exercise as possible? And is there a particular objection to our believing things that we are glad to believe, though there is no such objection to our eating things that we are glad to eat or taking exercise that we enjoy taking? What are the rational principles which dictate this kind of asceticism? Are they really necessary for our integrity?

REASON AS DISINFECTANT

Underlying this puritanism is the idea that the function of reason is essentially negative, critical, sceptical, sanitary. Thought exists purely to save us from error, as medicine exists to save us from disease. Particularly – for the moral drama is a very pervasive one – it exists to save us from indulgence in error.

Now thought certainly does have this negative, destructive, sanitary function. Thought is indeed – among other things – a kind of disinfectant, used, in accordance with a germ theory of error, to get rid of our sins and mistakes. But that could not possibly be its primary function, let alone its sole one. The creative, constructive work by which thought puts together our world-picture in the first place has to come first.

We have already encountered the weaknesses of this purely negative notion earlier. We did so first in noticing that the special glory of science could not possibly consist simply in avoiding being wrong, but must involve also being right about matters that were in themselves important. The point came up again in discussing the current idea that the work of science consists simply in attacking ready-made hypotheses by disproving them experimentally. This idea is hollow because it does nothing to explain where the hypotheses come from. It embodies a narrow and unreal idea of the general function of reasoning. It separates reason from the constructive imagination, which it treats as a kind of mysterious non-rational force or perhaps just a kind of plant that, by good luck, happens to produce hypotheses instead of tomatoes or cucumbers.

I am inclined to become rather sharper on the matter than we were earlier, and to add that this view is not just mistaken but is also a piece of bad faith. It is not something which could be seriously believed by anyone who attended to the way thought works. It is something which is recited and welcomed because it saves those who accept it from the responsibility for thinking critically about the conceptual schemes that they are using.

WHO IS THE ENEMY?

These narrow conceptions of scientific reasoning, are, however, just minor descendants of a much wider tradition which has had some very honourable members – the tradition of reason as embattled. There has often been a need to fight under its banner, and that banner, like the banner of freedom, has been raised against a bewildering variety of opponents. Sometimes there seems to be very little in common between these various campaigns except the need for opposition. In its most public and advertised forms, reason does very often have to be engaged on destruction.

What, then, are the things that reason needs to destroy? At its simplest, reason is seen as simply resistance to *any* error – attacking ignorance, confusion, inconsistency, superstition and false propaganda of all kinds. In that role everybody claims to support it. But since people are not unanimous about what they count as error, the ambiguities of the word 'rationalism' still echo these many different wars. The Rationalist Press Association is predominantly anti-religious. But within religion, rationalists who emphasize the role of reason contend against other people who stress faith or feeling or tradition or direct experience. Within philosophy, again, Rationalists who put their confidence in reason as a path to knowledge contend against Empiricists who put theirs in direct experience. And so on.

As these last two cases show, it is not possible to typecast this kind of conflict. Where 'rationalist' means 'atheist', it may also be likely (in our tradition) to mean innovator, rebel and sceptic. But within religion and within enquiries about knowledge, this is not so at all; the roles can sometimes be reversed. Rationalism is the older tradition. Battles fought in the cause of reason are in this very like battles fought in the cause of freedom; the opponent is always changing. Once the original, obvious enemy is defeated, the two opposed sides at once begin to melt and to form new patterns. Disregarded minorities show up and principles have to be re-examined.

THE FREQUENT IRRELEVANCE OF ATHEISM

This clarifying process has, I think, repeatedly been held up and distorted by a persistent impression that 'rationalist' means simply 'anti-God', a purely negative position, demanding no positive explanation. This idea is still very common. Peter Atkins, for instance, introduces his book *The Creation* by describing it as 'an essay in extreme reductionism, *and militant rationalism*',[7] meaning that his theme is to prove that there is 'no need to invoke the idea of a Supreme Being' to account for the world.

Atkins's book is excitably written, as if its author expected the Spanish Inquisition to appear at any minute if he wrote a disrespectful sentence about the First Cause Argument. It is not rationalistic in the sense of containing a lot of reasoning. It relies chiefly on colourful metaphors, designed to exalt the idea of Chaos into a creative force, a kind of substitute God. This move is only rationalistic in the sense that it is designed to make atheism attractive.

The trouble about this twist in the notion of rationalism is not just that there are other, quite different disputes to which the defence of reason is relevant. It is that anti-God arguments tend to cast reason always in a negative role. Through them, the idea caught on that it was usually more rational not to believe things, or at least, not things that people other than scientists believe. *Disbelief, as such*, acquired the reputation of being always more sensible than belief.

If one means by scepticism not just a cautious and enquiring temper, but a bias towards dogmatic denial, something rather odd is happening. This negative bias rests, I think, on what might be called the Hygienic View of Truth, the view that both Plato and Hobbes relied on in their attempts to purify the notions of (respectively) Soul and Body. On this view, truth is always there before us, but, like a statue bought in a junk-shop, it has become encrusted on the outside with error. We have only to peel the layers of error away for the underlying reality to emerge and be plain beyond question. And it is only these outside layers about which we have to be sceptical – the object itself, when it emerges, will be all right.

THE DANGERS OF FEELING

The simplest form of this conception is that the dirt consists in opinions induced by feeling, so that 'being scientific' means above all resisting feeling, which is the true enemy of reason. Thus C.H. Waddington ruled that 'The scientific attitude consists in the overruling of the more obvious emotions which might interfere with the unbiased appraisal of the situation.'[8] There is supposed to be no special difficulty in identifying these particular layers of feeling; they will stand out from the hidden statue as black from white. The only difficulty lies in our weak and self-indulgent attachment to them.

This line, of course, can only be held by people who have *not* taken up Hume's view that belief is just a form of feeling anyway. But Hume himself, in spite of his official championship of feeling, gave a lot of impetus to this particular drama by his moving description of the agonies suffered by sceptical thinkers like himself when they have to override their feelings in the service of their integrity.

Probably the best-known and most telling passage of his *Treatise* is his description of the chilling effect of maintaining his sceptical position among an unsympathetic public, and the heroic effort needed to resist temptations to relax it:

> I am at first affrighted and confounded with that forlorn solitude in which I am placed by my philosophy, and fancy myself some strange uncouth monster, who, not being able to mingle and unite in society, has been expelled all human commerce and left utterly abandoned and desolate. *Fain would I run into the crowd for shelter and warmth, but cannot prevail with myself to mix with such deformity.*[9]

There are, however, obvious consolations here which he does not mention. The idea of being so much nobler than everybody else is always cheering; isolation from the vulgar herd guarantees superiority. With the same satisfaction, Jacques Monod, referring especially to our weakness for ideas that attribute any sort of purpose to nature, contrasts the heroic stance of intellectual ascetics like himself with other people's inert conventionalism:

> We understand why so many thousand years passed before the appearance, in the realm of ideas, of those presenting objective knowledge as the only source of real truth. Cold and austere, proposing no explanation but imposing an ascetic renunciation of all other spiritual fare, this idea could not allay anxiety; it aggravated it instead ... It ended the ancient animist covenant between man and nature, leaving nothing in the place of that precious bond but an anxious quest in a world of icy solitude. With nothing to recommend it but a certain puritan arrogance, how could such an idea be accepted? If it is true, as I believe, that the fear of solitude and the need for a complete and binding explanation are inborn, ... can one imagine such an austere, abstract, proud ethic calming that fear, satisfying that need?[10]

Puritan arrogance, however, has its own appeal. It is very

convenient for virile posturing, well in tune with the spirit of the age, and it imposes no real costs where it is combined with compensatory faiths such as the ones we have noticed. Low-temperature rhetoric proves nothing; people are proud of their ability to withstand this kind of cold. It is not clear why Monod thinks the temptations of arrogance are necessarily any weaker or any less misleading than those of intellectual coherence. His Existentialist rhetoric is, however, well designed to prevent any such question from arising.

13

THE HUNGER FOR SYNTHESIS

FREEDOM TO CONFORM

Monod is not just a piece of history. His remarkable attempt to let science have its cake and eat it is still favoured today. It is not surprising that this combination made *Chance and Necessity* a best-seller and gave it great influence among scientists. It was one of the last really popular efforts to establish science as the true source of values. Its success no doubt owed much to Monod's making no effort to explain the discrepancy, relying instead on the dramatic rhetoric of open paradox – on simply saying a thing and then saying its opposite – a habit already common in Existentialist writings.

Monod could also, however, call on something more substantial lying behind the rhetoric, namely a faith. Existentialism is a credulous exultant faith in the human will as omnipotent, admirable and in effect an object of worship. Particular values or ideals are then exalted as supreme by the claim that they are the only values freely chosen by that will. The ideal that most commonly gets this arbitrary treatment is freedom itself, but Monod simply transferred the prize to knowledge, naming it, out of the blue, as the only end to be valued. All the other 'highest human qualities', such as 'courage, altruism, generosity, creative ambition' and the rest, are (he wrote) only means to knowledge.[1] If we ask why, we are simply told that this end has been chosen:

> In the ethic of knowledge, *it is the ethical choice of a primary value* that is the foundation ... The ethic of knowledge does not impose itself on man; *on the contrary, it is he who*

imposes it on himself, making it the *axiomatic* condition of authenticity for all discourse and all action ... [All the same, no other choice is really available, because] ... The ethic of knowledge that created the modern world is the only ethic compatible with it, the only one capable, once understood and accepted, of guiding its evolution.[2] (Emphases are Monod's)

HISTORICAL PERSPECTIVE

In this twenty-year-old formulation, then, there is no serious attempt to reconcile the two divergent claims about the status of science. I know of no more recent attempt to do it, and the intellectual climate has of course been getting steadily more hostile to the project. Officially, science grows ever narrower, ever more specialized, ever less willing to say anything about its relation to the world outside its journals. Even Existentialism, on which Monod relied, was already something of a bygone intellectual fashion at the time when he invoked it.

Many people, however, still search confusedly for a language in which to express the idea of science as a general saviour. The euphoric fantasies of some scientists about an endless, dazzling human future, to which we will come back shortly, are surely one symptom of such a hope. Another is the continued use for this purpose of moral idioms – such as the Existentialist one – which have in general been somewhat discredited, but are still called on because they are wide and obscure enough to do the job. Similarly, believers in the 'omnicompetence of science', especially in the United States, often still talk in a style extremely reminiscent of H.G. Wells.

In this sort of situation, where a lot of people are still trying to say things which the academics consider no longer defensible, it is often worth while going back to the more open statements that were made before that inhibition set in. Between the wars, it was widely believed that Marxism could solve the whole problem. We now stand at a point where it is becoming quite hard to see how anybody can ever have thought this. For that very reason, I think it is important for us to make that effort.

THE LURE OF THE ARTICULATE

Why did Marxism exert such a strong fascination on bright and learned intellectuals at that time? In particular, why was it so strongly welcomed as 'scientific' by people to whom that was an important recommendation, people who would have been suspicious and hostile towards anything that they had recognized as primarily a faith or creed? Why, for instance, did even people who had plenty of detailed objections to bring against Marxist theory, treat it with a respect that reads so oddly now?

For any one who lived through this epoch it is very striking to remember how highly Marxism used to score with an immense variety of sophisticated people. The notion of what is 'scientific' was then considerably wider. It could easily be used for large-scale, highly articulate theories on any subject-matter, though, for professional scientists themselves, an explicit link with physical science did increase the appeal. Engels had taken some trouble to supply this link for Marxism.

Even apart from this, though, the mere fact that Marxism was an elaborate, articulate theory impressed intellectuals extraordinarily. It often seduced even those who were both critical by nature and also disillusioned on many matters by long experience, like Bernard Shaw and the Webbs, and also scientists like Bernal and Haldane. It blinded them, both to gross faults in the theory itself, and to the huge gap between the whole mass of theory and the actual grim facts about the USSR. This experience has undoubtedly contributed to producing a widespread disillusion with theory in the West today, a philistine distrust of all thinking on general subjects.

THE UNDERLYING DRAMA

What made this acceptance possible, however, was surely the fact that Marxism was not only a theory but a faith, and in some ways a highly dramatic faith. Intellectuals are no more immune than other people to this kind of attraction, provided the packaging is one that they can accept. Once accepted, Marxism was emotionally very sustaining. It provided, in a most striking form, the promise of a better future, and it showed that future as approaching through a vast conflict of a kind that has tremendous emotional appeal. (Compare the lasting popularity of the Book of Revelation.)

This stress on conflict is the element which has made Marxism lastingly popular and nutritious in countries whose population is genuinely oppressed, and has no option but to fight its ruling class. Here it can strengthen idealism, making possible struggles against spiritual wickedness in high places which might have proved too discouraging without it.

This same stress on conflict is, however, also what has made it most deadly when it is professed by governments that are already in power. It gives such people a general excuse, allowing them to combine a sense of immediate emergency, justifying any sort of means against their opponents, with the confidence that their success in establishing the general happiness will vindicate these methods in the end. Notoriously, it blots out the extent to which, on a political scale, the means used will determine the end attained. This distortion of the relation between means and ends is important in understanding the fantasies that we shall have to consider.

Among intellectuals, Marxism attracted people who like the heroic because of its emphasis on conflict, and it reassured those among them who might have distrusted its purely emotional appeal by the cragginess of its texts. (At this level, it pays to be unintelligible. Ex-party members who have had to study the works, not just of Marx, Engels and Lenin but also of Stalin, can still testify to the stiffness of the ordeal.) For a time, this body of theory seemed to many thinkers to open an intellectual new Jerusalem, not just because it promised a millennium gained by conflict, but because it seemed to back this promise with a scientific status. It seemed like a means of extending the reliability of science over the whole area of practical thinking – a way of spreading it that would be free from doubtful value-judgments, since the theory was impartial, non-sectarian, essentially scientific. The modesty of science was to be combined with the constructive achievement of a new and central moral insight.

This hope appealed to the architectonic intelligence in many bright scientists. It satisfied that urge towards a general, comprehensive understanding which had brought them into science in the first place. It balanced the fragmentation of their specialized studies, allowing them to relate scientific aims to a wider humanitarian idealism. This was not a trifling gain; it was not a luxury. If we find no new way of making that relation – if nothing better now replaces Marxism – the loss will be serious. We are not in a position just to dance on the grave of Marx. We need to learn from his failures.

THE SCIENTISTIC PROJECT

The ideal of science which made Marxism seem appropriate has not changed. The hope at that time was that the demands of heroic Enlightenment puritanism were at last being met. A body of thought which was scientific because officially it owed nothing to feeling was available to save the world. To follow it simply involved subordinating all other ways of thinking – notably ethics and the other 'humanities' – to science. This position was not really that of Marx, who was primarily a historian, and who spent much more of his time discussing metaphysics, economics and politics than attending to physical science. Nevertheless, he had made this claim to scientific status which was growing so important in his day, and, as the idea of science narrowed, Engels had taken great trouble to endorse it.

Thus there emerged the idea that the synthesis which intelligent scientists hungered for could be found by starting from their own end – that physical science was, so to speak, always at the bottom, supplying the foundations on which alone other kinds of thought could be built. This gravitational metaphor is, as we have noted earlier, very strong and persistent. Descartes used it, putting physics at the bottom of the pile, and the picture of ethics and literature as something up in the air – something added perhaps as a pleasing, optional roof-garden – still possesses many people.

A moment's thought can show up its hollowness. Why would we do physics at all unless we had standards by which we judge it to be important? How would we choose between these standards if we did not know how to think morally? How, again, would we form such standards in the first place if our feelings and our imaginations had not been educated by many serious comparisons? Most of these comparisons must be presented by others who have thought seriously about them, so they take the form of history and literature.

And so on. Of course the point is not that the humanities supply the real foundation, but that the whole gravitational metaphor is wrong. Reasoning proceeds in all directions. The kind of 'support' that any given idea needs depends on the kind of doubt that is being raised about it at the time. If the doubt is a moral one, then the considerations it immediately needs are moral ones. Till they have been worked out, all the physical discoveries in the world will not help it. Of course, after the moral aspect has been brought in focus, those discoveries may happen to be relevant to solving the problem. But the idea that they are the universal, all-sufficient starting-point is naive and empty.

That idea, however, is just as prevalent in much Western thought today as it ever was in Marxism. It certainly formed one main strand in the conversion of considerable scientists such as Haldane and Bernal. The other strand was their credulity about the USSR, working on their quite genuine indignation against the oppression of the poor.

Here again, with hindsight, it is easy to see the effect of idealism acting on strongly combative temperaments and leading people who were fiercely – and often rightly – critical of their own society to be childishly uncritical of a distant one, even when they visited it often. To resist this bias called for a very firm philosophical balance, which, among this group, perhaps only Joseph Needham possessed. The others, clever and learned though they were, were unaware that they were indulging in romantic projection of their desires on to remote and unknown scenes as an escape from immediate and disagreable problems. This kind of unawareness, unfortunately, is still with us, and it forms an important element in the future-fantasies we have to consider.

If it seems to us now that these people were exceptionally naive, we should perhaps remember that it is always easy to see the errors of one's forebears. The level of public credulity does not really change much. People are always straining at gnats and swallowing camels; the main change is in the particular camels that will go down at any given time, and even the nature of those camels does not always change. When the Soviet government sacked Vavilov and his school of highly effective agronomists in 1948 to replace them by Lysenko's set of green-fingered charlatans, it gave exactly the same reasons for doing so that have been given recently for closing so many university departments – Vavilov was not getting practical results fast enough. The future of Russian agriculture was indeed sacrificed at that time to abstract economic and political theory. But it was certainly not the last thing to be so sacrificed.

THE CHOICE OF FAITHS

Those who accepted Marxism between the wars mainly thought that they were doing so because they were austere, realistic, modern people who had outgrown the consolations of religion, and were responding to purely rational, scientific arguments. If we now reject those arguments, we shall probably conclude that they did it because they needed a faith. But why *this* faith?

It is worth while looking at the others that were available. Seekers for faith tended to find Marxism more nutritious than Freudianism, partly because it was much more optimistic – it really did promise a better future – and partly because it was larger, more comprehensive, more unifying. Marxism had something positive to say and recommend about almost every aspect of human activity, whereas Freudian thinking concentrated on the inner problems of the individual.

People did, of course, turn to Freudian thinking for their personal salvation, as they still do. But in general this had to be a private salvation, one which involved to some extent turning one's back on the problems of an irredeemable world, rather like Gnostics and Manichees in an earlier age. This therapeutic refuge is of course still available, and in the United States it is now a very important element in shaping people's conception of salvation. But to make it a central element demands a depth of individualism which is rarer on this side of the Atlantic – perhaps in the end a kind of moral solipsism.

This extreme individualism is indeed itself also a possible faith, and one for which there has been a good deal of propaganda in this century – propaganda which has roots in both Nietzsche and Social Darwinism, and prophets as various as Sartre, Ayn Rand and the sociobiologists. In another fifty years, writers looking back and wondering how their parents can possibly have been so foolish will perhaps be heard marvelling at the romantic excesses of twentieth-century individualism, and noting how its economic expression in monetarism and exaltation of market forces attracted just the same kind of earnest intellectuals who had earlier been converts to Marxism.

In the twenties and thirties, however, monetarism was not available. The Eastern religions were not widely considered either, though indeed Haldane – always ahead of his time – did settle in India after the Lysenko affair had disillusioned him with Stalinism, and became seriously interested in Hinduism, which had always attracted him.

The faith that, along with Marxism, was most attractive to British intellectuals at that time was Roman Catholicism, which also offered fairly tough intellectual argumentation, was also applicable to every aspect of life and, of course, also proposed an eventual happy ending. (Fascism, which was a much less wide-ranging theory, did not usually become attractive as a faith unless it was combined either with Roman Catholicism or with mysticism about race.) But the distinguished intellectual converts to Rome tended to be people from the humanities – Edith Sitwell, Evelyn Waugh, Graham Greene. For people who put their faith in science, Marxism was the obvious option, a challenge which they had to meet if they became involved in general reflection at all. You had, as it were, to show reason why you were *not* a Marxist.

The only serious competitor with Marxism for this role was evolutionism – the Lamarckian belief in a vast escalator-process by which the human race has been brought to the top of the world's animal populations and will be carried on securely to an indefinite series of further glories in the future. The reason why Waddington could view Marxism with a certain detachment was that he – like Julian Huxley – was deeply committed to evolutionism, and believed it to be wholly scientific.

This belief is itself strange, because, officially, the current Darwinian view does not see evolution as an escalator, but as a sinuous, branching, radiating pattern – not a staircase but perhaps a bush or a seaweed. Life-forms diverge from each other to meet particular needs in their various environments. Our own species figures then only as one among the many, with no special status or guarantee of supremacy. This notion has, however, always been found far less exciting than the escalator model, which has been enormously popular ever since it was promoted by Herbert Spencer, in spite of Darwin's own rejection of it and its evident complete irrelevance to his theory.[3]

ESCALATOROLOGY

This belief in an endless evolutionary escalator exalting the human race, which is often seen as part of science, is a prime example of the dreams, dramas, myths or fantasies out of which faiths are constructed to fill the vacuum which is left when more familiar ones are abandoned. This process in itself ought not to surprise us. In any profession or sect, seductive but irrelevant ideas do get passed round and added to the official core as easily as pheromones in an ants' nest.

For instance, it might be said that the Christian church, too, early acquired some ideas that would probably have much surprised its founder – among others, a special objection to sexual activity, an approval of war ('crusades'), and a strong identification with the forces of worldly government. To later generations, these things have often looked essentially Christian. The only reason why we expect this kind of thing not to happen to 'the Church Scientific', as T.H. Huxley called it, is that science has loudly and publicly forsworn them. But it is rather simple-minded to put one's trust in such forswearings.

14

EVOLUTION AND THE APOTHEOSIS OF MAN

LOOKING AHEAD

What, then, are these alarming quasi-scientific dreams and prophecies? They are, as we have seen, predictions of the indefinitely increasing future glory of the human race and perhaps its immortality. They claim scientific status, not just because they appear in scientific books, but also because they claim to take their start from the biological Theory of Evolution. They lay great stress on this theory, although they use a conception of it quite different from the one that is officially taught as part of modern biology.

Evolution, in these prophecies, figures as a single, continuous linear process of improvement. In the more modest form in which some biologists have used it, this process was confined to the development of life-forms on this planet. But it is now increasingly often extended to do something much vaster – to cover the whole development of the universe from the Big Bang onward to the end of time – a change of scale that would be quite unthinkable if serious biological notions of evolution were operating. To ensure that this scheme works, physicists have added further elements to the story, such as the vision of the Omega Point which we have already noticed in connexion with the Anthropic Principle, and others which we will discuss presently.

This supposed vast linear process – which we might call hyper-evolution – will, it is claimed, take us away from our present way of life altogether, usually by removing us from this planet. Space travel, genetic engineering, artificial intelligence and other vast but vague cultural changes will form part of it.

147

Yet, it is still supposed to be a single, continuous, infallible process, and reliance on the whole still rests on the original theory of evolution, which is held to predict it all.

FROM THE EARTH TO THE STARS

The more modest, purely terrestrial form of this project is laid out in the last chapter of an otherwise normal scientific book about the origins of life by the molecular biologist William Day. (Final chapters, incidentally, are the main habitat of this shy literary life-form. It tends to be cautious of appearing in the main argument of books.) Turning from the remote past to a more exciting future, Day explains that:

> He [man] will splinter into types of humans with differing mental faculties that will lead to diversification and sepa-rate species. From among these types a new species, Omega Man, will emerge either alone, in union with others, or with mechanical amplifications to transcend to new dimensions of time and space beyond our comprehension – as much beyond our imagination as our world was to the emerging eucaryotes ... If evolution is to proceed through the line of man to a next higher form, there must exist within man's nature the making of Omega Man ... Omega Man's comprehension and participation in the dimensions of the supernatural is what man yearns for himself but cannot have. It is reasonable to assume that man's intellect is not the ultimate, but merely represents a stage intermediate between the primates and Omega Man. What comprehension and powers over nature Omega Man will command can only be suggested by man's image of the supernatural.[1]

These prospects might seem both odd enough and grand enough for many of us, but they have one drawback: they are not permanent. The Second Law of Thermodynamics is held to ensure that some day the success of the human race will end, and this is found intolerable. Without permanence (said Stephen Weinberg in his own last chapter) 'the more the universe seems comprehensible, the more it also seems pointless'.

Instead of asking why Weinberg took the meaning of human

life to depend wholly on its going on for ever, his colleagues therefore looked for ways of proving that in fact it *will* go on for ever. Like the fisherman's wife in the fairy-tale, who was never content but kept sending her husband back to ask the magic fish for something more, Freeman Dyson refused to bow to the Second Law of Thermodynamics and put in instead for species-immortality.

DYSON TO THE RESCUE

Dyson answers Weinberg by arguing that the Second Law of Thermodynamics need not have this distressing effect. The universe may not, after all, ever run down completely. Though things will indeed largely even out at a low temperature, though matter as we know it will mostly vanish, this process may (he argues) stop short of its absolute limit. In scientific calculations meant to establish this, he makes a number of large assumptions which he explicitly says are wishful thinking. He puts this admission in elegant terms such as 'My answers are perhaps only a reflection of my optimistic philosophical bias';[2] 'since I am a philosophical optimist, I assume . . .' and 'On grounds of simplicity I disregard these possibilities and assume the proton to be absolutely stable' (which it needs to be if his plans are to work). And again, 'I shall not discuss the closed universe in detail since it gives me a feeling of claustrophobia to imagine our whole existence confined within this box.'[3]

But there is nothing philosophical about wishful thinking, and it is not supposed to be part of science.

Dyson concludes, however, that something or other will still go on. If we want to survive therefore, we need only adapt ourselves so as to inhabit that something. This, he writes, establishes a universe

> very different from the universe which Steven Weinberg had in mind . . . I have found a universe growing without limit in richness and complexity, a universe of life surviv-ing for ever and making itself known to its neighbours across unimaginable gulfs of space and time. Is Weinberg's universe or mine closer to the truth? One day, before long, we shall know.[4]

Somewhat surprisingly, this seems to place the project well

within the range of current enquiry. How is it going to work?

> The energy resources of a galaxy would be sufficient to support indefinitely a society with a complexity about 10 times greater than our own . . . No matter how far we go into the future, there will always be new things happening, new information coming in, new worlds to explore, a constantly expending domain of life, consciousness and memory.[5]

This conclusion, with its remarkable figure of 10, is calculated by using a quantity Q which 'is a measure of the complexity of the molecular structures involved in a single act of human awareness'. Since nobody has the slightest idea how to isolate, measure or count such single acts, this does not mean very much. As we examine the expected conditions further, we shall have a chance to decide how well they suit the description of 'a constantly expanding domain of life, consciousness and memory'.

We ask next who is going to be there to enjoy all this? It certainly will not literally be ourselves. Will it be anybody or anything at all like us? Well, perhaps not exactly. Dyson goes on:

> It is impossible to set any limit to the variety of physical forms that life may assume . . . It is conceivable that in another 10^{10} years, life could evolve away from flesh and blood and become embodied in an interstellar black cloud (Hoyle 1957) or in a sentient computer (Capek 1923).[6]

In fact, his arguments appear to leave only the one possibility of a sentient computer constructed on the model of the Black Cloud in Hoyle's story. For, as he explains, on the view he thinks most probable, 'human-sized objects will disappear . . . but dust-grains with diameter less than about 100μ [i.e. 100 microns, about 250th of an inch], will last for ever'. Accordingly,

> The preferred environment for life in the remote future must be something like Hoyle's black cloud . . . We cannot imagine in detail how such a cloud could maintain the

state of dynamic equilibrium that we call life. But we also could not have imagined the architecture of a living cell of protoplasm if we had never seen one.

This, it should be noticed, is an all-purpose excuse for never explaining anything. How many features of the universe could we have invented if we had never seen them? This consideration does not make the proposition that human consciousness can be transferred to a stellar dust-cloud any more plausible than the proposition that it can be transferred to a chair. (A further argument of a kind is given, but it will be best to consider it with Barrow and Tipler's version, in the next chapter.) Dyson, however, is satisfied that he has shown, not just that such beings can exist, that they can take on human consciousness and that they can survive indefinitely, but also that they can, at will, dominate their environment to such an extent that they might be able, if necessary, to 'break open a closed universe and change the topology of space-time so that only a part of it would collapse and another part of it would expand forever'.[7] He accordingly concludes:

Whether the details of my calculations are correct or not, I think I have shown that there are good scientific reasons for taking seriously the possibility that life and intelligence can succeed in *molding this universe of ours to their own purposes.*[8]

This cheerful estimate is not disturbed by the sharp limitations on activity which he says are necessary in order to function at all at such a low temperature – limitations such as the need for great slowness in all processes, and for conserving strength by frequent hibernation. 'Life keeps in step with the limit on radiated power by lowering its duty-cycle in proportion to its temperature.' And, as he explains,

I have not addressed at all the multitude of questions that arise as soon as one tries to imagine in detail the architecture of a form of life adapted to extremely low temperatures. Do there exist functional equivalents in low-temperature systems for muscle, nerve, hand, voice, eye, ear, brain, and memory? I have no answer to these questions.

151

BERNAL AND THE ORIGINAL DREAM OF DISCARNATION

Such questions, however, seem rather important for considering the desirability of the scheme. If the answer to questions about the senses is 'no', then it sounds as if the beings will need all to be Boddhisattvas, entirely engaged in contemplation with no dependence on outside perception. If so, then their spiritual training ought surely to be a central part of the scheme. This moral and social dimension needs sharp attention. In these discussions it is never mentioned. Neither is the question of what relation we – here and now – are supposed to have to these beings. Why should their fate concern us at all? Obviously, beings capable of existing in this way would have to be very unlike us indeed. They are not physically descended from us, and most of our culture could not possibly be transmitted to them. Their physical nature, however, is the only topic that interests these theorists. To explain it, Dyson enthusiastically quotes a passage from Bernal's Utopian prophecy of 1929, *The World, the Flesh and the Devil*:

One may picture, then, these beings, nuclearly resident, so to speak, in a relatively small set of metal units, each utilizing the bare minimum of energy, connected together by a complex of aetherial intercommunication, and spreading themselves over immense areas and periods of time by means of inert sense-organs, which, like the field of their active operations, would be, in general, at a great distance from themselves. As the scene of life would be more the cold emptiness of space than the warm dense atmosphere of the planets, the advantage of containing no organic material at all, so as to be independent of both these conditions, would be increasingly felt . . . The new life would be more plastic, more controllable, and at the same time more variable and more permanent than that produced by the triumphant opportunism of nature. Bit by bit the heritage in the direct line of mankind – the heritage of the original life emerging on the face of the world – would dwindle, and in the end disappear effectively, being preserved perhaps as some curious relic, while the new life which conserves none of the substance and all the spirit of the old would take its place and continue its development ... Consciousness itself may end or vanish in a humanity that has become completely aetherialized, losing the close-knit organism, becoming masses of atoms in space communicating by radiation, and ultimately resolving itself entirely into light.

All this is assumed to be both desirable and also biologically possible. (If biologists or humanists speculated like this about physical possibilities, might there not be trouble?) Here is a more recent version, one which takes advantage of new techniques to avoid biology altogether, but does make a hasty gesture towards the moral aspect:

It is, however, possible that we could program computers to be not only more intelligent, but in other ways better than ourselves; kinder, more loyal, more unselfish, without deceit, and in fact more perfectly the possessors of man's most valued qualities than any men have yet been able to be ... Furthermore, freedom from the biological ball and chain would allow the new beings freedoms hitherto only dreamed of or imagined as the most improbable fantasy. Since the nature of the individual would be that of a logical organization, a program which merely had to inhabit a computer in order to actualize itself, it could be transmitted through space with no more difficulty than a television program, and at the speed of light, to some suitable recipient machine. Furthermore, since the actual hardware of the machine would be replaceable, without destroying the program that specifies its interconnexions, an individual would be effectively immortal, until it chose to replace itself with a better version. Space travel, between solar systems which for reasons of mere distance seem impossible for biological man, immediately begins to seem plausible.[9]

THE QUEST FOR POWER

These ideas are not just an inconsequent private aberration of certain authors. They have a public meaning; they are a tip which really does have an iceberg. Their intense tone, combined with the professional authority of the writers, hammers their message through to a much wider public than would normally read or remember or understand books about science. They are not just ritual. They form part of a peculiar culture, apparently scientific in its subject matter but highly emotive in its tone, which links the more solemn areas of science-fiction with the more mythical aspects of popular science. So it is not a waste of time for us to ask, what is going on here? What is the message of these ambitious visions?

In the first place, of course, there is here a power fantasy. 'We' are going to become supreme in the universe. This startling claim is often not explained; it tends to appear out of the blue without real roots in the course of the argument. Thus, Dyson's prediction that the human race will succeed in 'molding this universe of ours to their own purposes' emerges in his discussion as a sudden piece of wish-fulfilment, quite without support from anything else that he says. Barrow and Tipler's startling claim that life – that is, ourselves – 'will have gained control of all matter and forces' seems an equally arbitrary addition to the acquiring of information, which is their official program. Similarly Day gives no reason for supposing that 'what comprehension and powers over nature Omega Man will command can only be suggested by man's image of the supernatural'.

Bernal's remark that the new life would be 'more plastic, more controllable' than the present one expresses this same eagerness for control and distaste for what is uncontrolled or natural. This theme is central to his book. Throughout, he regards any form of human dependence as disgraceful parasitism, not (apparently) noticing that such dependence is unavoidable. As we have seen, he takes a very firm line with the stars:

> Man will not ultimately be content to be parasitic on the stars but will invade them and organise them for his own purposes . . . If energy is still needed, the stars cannot be allowed to continue in their old way, but will be turned into efficient heat-engines.[10]

And again, with regard to natural scenery:

> The art of the future will need an infinitely stronger

154

formative impulse than it does now. The cardinal tendency of progress is the replacement of an indifferent chance environment by a deliberately created one. As time goes on, the acceptance, the appreciation, even the understanding of nature, will be less and less needed. In its place will come the need to determine the desirable form of the humanly controlled universe which is nothing more nor less than art.

ESCAPE FROM FEELING AND FROM THE BODY

Bernal does, however, show a little more interest than his successors in questions of *why* these extraordinary developments must be expected or considered necessary. Noting, in Freudian style, that some people will ask whether science itself is actually a necessary human aim or just some kind of 'perversion' of our natural motives, Bernal replies that we must no longer suppose that our natural motives are good enough to guide us. Indeed, it is of great importance for human progress that we should not be led by our feelings. In the improved future, the physiological balance determining emotion

> will not be, as in us, at the mercy of the uncontrolled interactions of individual and environment. Feeling, or at any rate feeling-tones, will almost certainly be under conscious control; a feeling-tone will be induced in order to favour the performance of a particular kind of operation.[11]

There is a significant muddle here. Bernal has not noticed that feeling shapes choice long before self-control operates, that all motives are feelings. In deciding (for instance) to take up space-travel – or indeed to do science in the first place – we must be led by some desires, some aims already favoured and accepted by our feelings as well as our intellects. Bernal takes no interest in what such motives might be. In answer to the doubts of an objector who might ask what space-colonization is for, he replies that it is required by our need for unrepressed activity. What is essential for a healthy human life is, he says, that our speculative thinking should not remain mere fantasy, but should be expressed in action:

> A sound intellectual humanity will never be content with

repeating itself in circles of metaphysical thinking like Shaw's Immortals, but will need a real externalization in the transforming of the universe and itself.[12]

Here Bernal suggests that merely to sit and do science would be to fantasize, and though he would no doubt prefer this to the more circular fantasizing of metaphysics, he still insists that it cannot hold a candle to real action. It plainly does not strike him that the planning of immensely, inconceivably distant action is itself a fantasy occupation, quite detached from the practical considerations that would arise if anyone actually reached the point of choice.

There is something bizarre, too, about his assumption that all speculative thought is just a frustrated attempt at action – that straightforward curiosity which wants the truth for its own sake must be a 'perversion' rather than one natural motive among others. On Bernal's suggestion here, knowledge is not really a human aim at all. This is very odd indeed in view of the exaltation of knowledge which informs so much of these fantasies, including his own, and still odder in view of the contempt he expresses for the human body and its ordinary activities:

Modern mechanical and modern biochemical discoveries have rendered both the skeletal and metabolic functions of the body to a great extent useless ... Viewed from the standpoint of the mental activity by which he [man] increasingly lives, it is a highly inefficient way of keeping his mind working. In a civilised worker the limbs are mere parasites, demanding nine-tenths of the energy of the food and even a kind of blackmail in the exercise they need in order to prevent disease, while the body organs wear themselves out in supplying their requirements ... Sooner or later the useless parts of the body must be given more modern functions or dispensed with altogether.[13]

WHAT ARE POSSIBILITIES?

I am sorry if the reader is growing tired of oddities. I am not quoting them just because they are odd, but because they do seem to me significant. Much of this thinking makes little sense on a literal interpretation, but plenty of sense if read as wish-fulfilment, as a range of fantasies about power which are not meant to be realistic at all, yet are presented as factual. What makes any other interpretations so hard is the way in which these very well-informed scholars constantly ignore the accepted limits of physical possibility.

It is not enough for them to write – as they do from time to time – 'this certainly looks rather improbable; I am not guaranteeing that this will work; I am only exploring an exciting possibility'. Nor is it enough to complain of the timidity of other theorists. It is necessary to explain why one thinks this idea worth spending the reader's time on at all.

That pigs will fly is not just improbable; the skeletal structure of mammals makes it an impossibility. Angels, as traditionally represented in Western art, similarly are not anatomically possible. Again, a human child a year old cannot grow up into an elephant. Accordingly, a biologist who wished to canvass the project of genetically engineering angels or winged pigs or elephant-children would have to start by explaining fully how these objections had been met. The same thing is true of Dyson's offer (which we shall meet shortly) to adapt human beings for outdoor life on Mars. It is not enough, in such cases, for theorists to say that they find such ideas exciting, or that they have an optimistic temperament.

The fact that there are such limits – that possibility is not infinite – is often obscured today by the dramatic progress of certain selected kinds of thought and technology. Suggestions about limits tend to be brushed aside by the protest, 'Oh, but this is only in its infancy.' An infant grass-snake will, however, not grow up to be a spider, nor an acorn to be a pear-tree. Moreover, biology reminds us that most infants in most species never grow up anyway, that an individual which reaches maturity is highly exceptional. It also tells us that no organism goes on growing for ever, and that all of them die in the end.

This 'infancy' metaphor has been deeply misleading. Removing it, we note simply that there is no guarantee of reliable, linear progress for all our exciting thoughts and activities. Most human ideas which are tried at all prove useless or are soon superseded. Most which persist must keep changing their function and direction with the life around them. And all of them have their limits in the end.

Pointing out such limits honestly is a very important function of scientists. Of course their beliefs are corrigible; of course particular vetos may turn out mistaken. (Egg-laying mammals proved possible after all.) But to say that something is possible means much more than that it might happen if beliefs that currently look certain prove mistaken. It means that it is actually compatible with those beliefs. On that question, we do expect to be able to rely on the rough map that experts give us. We expect scientists to make it realistic.

Theorists constructing these space-colonization dreams do not attempt this. Until the idea of substituting computers for people came along, they did not just grossly exaggerate the ability of the human mind and body to survive harsh conditions; they talked as if that versatility were known to be unlimited. Now that computer substitutes have been proposed, they talk with equal, unconsidered confidence as if it were already known that human consciousness can be transferred into them without loss, and that they can be made to perform all important human functions. As we noted earlier, they also constantly ignore the huge difference of scale, both in time and space, between our own planetary affairs and those of the cosmos. This is the kind of mistake we ordinarily expect from scientifically illiterate or careless people, and from science-fiction writers who are allowed to colour it by using imaginary devices like 'hyperdrive'. When eminent scientists commit it, there has to be a reason.

Bernal set an appalling example here, observing, for instance, quite casually that 'our desires are . . . *already* tending to be the chief agent of change in the universe'[14] (Emphasis mine). Even cosmologists who are not trying to bolster a case for space-colonization, and whose work involves assessing the probability of such large-scale cosmic events, sometimes fall into similar language. Compare here a striking passage from Paul Davies's book *Superforce*:

> Letting imagination have free rein, it is possible to envisage mankind one day gaining control over the superforce. To achieve this would enable us to manipulate the greatest power in the universe, for the superforce is ultimately responsible for generating all forces and all physical structures. It is the fountainhead of all existence. With

the superforce unleashed, we could change the structure of space and time, tie our own knots in nothingness. and build matter to order. Controlling the superforce would enable us to construct and transmute particles at will, thus generating exotic forms of matter. We might even be able to manipulate the dimensions of space itself, creating bizarre artificial worlds with unimaginable properties. *Truly, we should be lords of the universe.*[15] (Emphasis mine)

What would it mean to control, or to unleash, such a natural force? Parts of the context suggest that Davies has in mind only changes produced momentarily on a very small scale, within an experimental apparatus. But if this is all, the grandiose language is simply ludicrous. If he does mean something wider, how could any of the drastic tricks he wants to play be done without annihilating our own bodies – or whatever substitute for our own bodies might by then have been invented? It is hard to see how passages like this can be understood other than as self-indulgent, uncontrolled power-fantasies.

POWER FOR WHAT?

Power, however, should have an expected use. What is all this power wanted *for*? It remains extraordinarily abstract. It seems to be wanted for its own sake rather than for any notion of some work for which we might need it. The idea that we want it in order to make the universe last longer is not very convincing, nor is it made central to these discussions. In fact, this vision of increased power, this Utopia – if it is a Utopia – is not much like other visions of future possible better worlds.

Ordinary Utopias have their own faults, but they usually serve quite a useful function in suggesting what is wrong with the world as it is at present. They are direction-indicators, general proposals for setting right particular things which they show up as wrong. That is the function of books like Plato's *Republic* and More's *Utopia*, and indeed of *Das Kapital*. But this quasi-scientific scheme seems to be more of an attempt to escape from these problems altogether. It presupposes that all social and political problems have been already solved. For instance, consider the political difficulties of setting Bernal's or Dyson's whole scheme in motion. How, for a start, do you secure agreement about turning everybody into light, or into stellar dust, or relatively small sets of metal units . . . ?

Haldane, much more clear-headed than most of his followers, saw what extraordinarily obsessive discipline and unanimous devotion would be needed to make any such transformation possible. He has his Venusian colonists explain the point:

> Among those whose descendants were destined for the conquest of Venus a tradition and an inheritable psychological disposition grew up such as had not been known on earth for twenty-five million years. The psychological types which had been common among the saints and soldiers of earlier history were revived. Confronted once more with an ideal as high as that of religion, but more rational, a task as concrete and infinitely greater than that of the patriot, man became once more capable of self-transcendence . . . The price (for such evolution) is paid by the individual, but the gain is to the race. Among ourselves, an individual may not consider his own interests a dozen times in his life. To our ancestors, fresh from the pursuit of individual happiness, the price must often have seemed too great, and in every generation many who have now left no descendants refused to pay it.[16]

These are the beings who propose to colonize the rest of space. Summing up his prophecies, Haldane added:

> I have pictured a human race on the earth absorbed in the pursuit of individual happiness; on Venus mere components of a monstrous ant-heap. My own ideal is naturally somewhere in between, and so is that of almost every human being alive today. But I see no reason why my ideals should be realised. In the language of religion, God's ways are not our ways. In that of science, human ideals are the product of natural processes which do not conform to them.[17]

BERNAL: THE SCIENTISTS TAKE OVER

Bernal, however, simply saw these problems as calling for a take-over of the leadership by scientists. Much, he explained, would already have been achieved by a Marxist revolution.

The state having withered away, most people would be living contentedly under the dictatorship of the proletariat, 'statically employed in leading an idyllic Melanesian existence of eating, drinking, friendliness, love-making, dancing and singing'[18] benignly ruled by the scientists, who would have found 'the means of directing the masses in harmless occupations under the appearance of perfect freedom'. Only then would the scientists themselves start to mechanize their own bodies and to explore space. Then, 'from one point of view the scientists would emerge as a new species and leave humanity behind; from another, humanity – the humanity that counts – might seem to change en bloc, leaving behind in a relatively primitive state those too stupid or too stubborn to change'.

The rest of the species, suitably impressed, would (he explains) probably follow this example. If, however, they were so misguided as to object to these changes, 'it would then be too late for them to do anything about it. Even if a wave of primitive obscurantism then swept the world clear of the heresy of science, science would already be on its way to the stars.' Should there still be further trouble, the scientists must take a firm line. 'There may not be room for both types in the same world and the old mechanism of extinction will come into play. The better organized beings will be obliged in self-defence to reduce the numbers of the others, until they are no longer seriously inconvenienced by them.'

Interestingly, these bold scenarios for conflict do not extend to the reconstituted scientists themselves. Their feelings will be better organized by means of an improved physiological balance. 'This balance will not be, as with us, at the mercy of the uncontrolled interaction of the individual and the environment. Feeling, or at any rate feeling-tones, will almost certainly be under conscious control; a feeling-tone will be induced to favour the performance of a particular kind of operation.'[19] So harmonious will they have become that, when twenty or thirty thousand of them live at close quarters in a hollowed-out asteroid, 'there would probably be no more need for government than in a modern hotel'.

Incidentally, in case we are worried about the practical problems of this asteroid, Bernal kindly tells us that 'owing to the absence of gravitation, its construction would not be a feat of any magnitude' (p. 24). This startling remark does not, evidently, flow from genuine ignorance of the drawbacks of life without gravitation, for a few pages later he remarks that 'Dust would be an unbearable nuisance and would have to be suppressed, because even wetting it would never make it settle.'

How, then, do you excavate an asteroid and live in it? This is just one of numberless examples of the calm, barefaced, dead-pan dishonesty which, when indulged in by those carrying the authority of science, so illicitly eggs on readers to take these schemes seriously. I am sorry if such words are in bad taste, but they are appropriate and the matter needs attention. This is dishonesty and not the symbolism of fiction because, as we have seen before, Bernal made it clear that his book was not intended as fiction or poetry but as literal truth.

No doubt the habit flows from self-deception. It shows the same romantic credulity about remote affairs, the same obstinate persistence in wish-fulfilment, which led Bernal to accept the Soviet government's assurances that all was well throughout the Lysenko scandal – even though he knew the people involved – and which later kept him defending Kremlin policies for twenty years longer than Haldane, right up to his last illness in 1969.[20] Bernal was a notable scientist, and in many ways an impressive man. But considered as a prophet – as a wise guide for the future – he is an impossible choice.

GETTING AWAY FROM THE ORGANIC

Whatever may be thought of Bernal's social and political analysis, today's space emigrants, still using and quoting his fantasies with unreserved approval, produce no improved substitute for his moral vision. Indeed, they hardly seem to see these difficulties at all. They are not interested in making our existing human life better, nor in understanding it, nor even in simply making it work, but in getting away from it. Their central motive does not seem to be any kind of reforming idealism, but simple fear.

In the first place, there is here surely the plain fear of death – ordinary, old-fashioned, individual death. The hope is that we can identify ourselves somehow with a set of future immortals. The pronoun 'we' is often relied on for this work. Thus Dyson writes, 'If it turns out that the universe is closed, we shall still have about 10^{10} years to explore the possibility of a technical fix that would burst it open.'[21] But beyond that, as often happens, the unbridled, unconfronted fear of death brings with it a fear of life as well, and particularly a fear of the body.

Life is seen as leading to death – as indeed it does – and escape has to begin by a retreat from life itself. Scared by the conflicts and complexities of our present animal existence, the visionary wants to exorcize the organic and the biological altogether. Thus Bernal's new beings have 'inert sense organs' which 'like the field of their active operations would be, in general, at a great distance from themselves' and end up by being 'totally aetherialized into light'[22] – a picture which would surely have been welcome to ascetics like Plato and Tertullian.

The implausibility of anthropic space-prophecies would not, of course, matter to the rest of us if it stood alone. Implausible predictions in academic subjects are two a penny; they normally concern only the fellow-specialists involved. What calls for general discussion is the meaning these predictions have for their authors and readers, more especially if these are influential people. What kind of considerations make up that meaning?

I have already mentioned two crude motives for this kind of prophecy – the fear of death and the lust for power. To mention crude motives may seem like bad taste, but it is unavoidable. In examining ideas which are in this way somewhat unofficial, ideas which are not being subjected to critical thinking, we can expect crude motives to be central, and what is central must not be ignored.

There is, however, obviously also something more respectable, or at least more positive. There is a genuine exaltation of the intellect. Day depicts his 'Omega Man' as essentially intellectual, indeed mainly engaged in doing science. Barrow and Tipler represent the mysterious advance of 'life' to occupy the universe, not just as a conquest of power, but as essentially an acquisition of knowledge.

Bernal simply took it for granted that any increase in attention to science at the expense of other human activities was good and desirable. He also thought it inevitable, because it was the obvious occupation for humans to turn to when – as would shortly happen – they had met all their physical needs. Alternative possible activities seemed to him plainly frivolous, self-indulgent or (as he revealingly puts it) 'Melanesian';[23] not important. He therefore speaks of the fully vaporized or aetherialized state which humanity is to reach as a 'new life, which conserves none of the substance and *all of the spirit* of the old'.

The claim is remarkable; what does it mean? Something more than the crude motives is certainly involved here. Both Bernal and his followers are surely in part moved by spiritual longings of a fairly traditional kind. They are really distressed by the contrast between the narrowness, meanness and brutality of much existing human life and the far better things of which humanity seems capable. They do not believe that we cannot do better, nor that (as the cruder forms of both Marxist and Capitalist thinking suggest) all we need to do so is more material provision. They want a nobler mental life, and it seems to them that their own occupation – science – must be the one to provide it.

Rejecting the narrow notions of its function which we noticed at the beginning of this book, they want science to provide salvation. But they want it to do this alone. The project therefore must be ambitious indeed. It must be able to promise glory and immortality reminiscent of the strongest offers available from religion, but more seductive still because they offer complete supremacy. They are not to be compromised by any unwelcome competition from God.

15

DYSON, ANIMISM AND THE NATURE OF MATTER

SHOULD RELIGION BE KEPT OUT OF SCIENCE?

In understanding this project, it will help us, I think, to start by looking seriously at Freeman Dyson's prophecies. Besides being a very distinguished physicist, Dyson is an imaginative person who has done a good deal to reshape the earlier Wellsian vision for his own age. Along with Bernal, he has originated most of the interesting mistakes that are now central to it, and they are very well worth examining.

He is not by any means simply the crude space-wizard that quotations so far might suggest. There are, in fact, two Dysons. Dyson 1 has been an active and very well-informed campaigner against the excesses of nuclear weaponry, and has of late become much concerned about protecting the environment. He firmly rejects Monod's crude, fatalistic drama about the rule of chance and the isolation of the human race – the drama that originally made the universe seem meaningless and so (as we have seen) made this manufacture of everlasting life seem necessary in the first place. Dyson 2, however, fully supports and develops Bernal's dream of colonizing space and converting the human body into non-solid forms. He has added the hope of making it last for ever. And, as we have seen, he is the pundit who is now hailed as having given that dream the secure authority of science.

Quite why a man who has demolished Monod's fatalism so well should still want to spend his ingenuity on unlikely schemes to defeat space and time is not easy to see. One reason surely is a deeply religious temperament, not easily satisfied about the destiny of the human soul. Another, however, does seem to be the excited pursuit of power for its own sake which we have just noticed. Dyson's history brings an important new element to this. He worked closely, after World War II, with the remarkable team of physicists in the USA who had been occupied in inventing the atomic bomb at Los Alamos.

As all accounts show, this extraordinary feat of intellectual co-operation generated a close bond and a huge sense of confidence among the scientists concerned. They felt omnipotent about technical fixes, about means. And – as Dyson himself points out in his autobiography – there had been absolutely nothing in their work to make them critical about aims. Apart from Oppenheimer and Dyson himself (who had lived through the war in Britain), they were not uneasy about the bomb. They had not even noticed that their achievement had been made easy by the fact that it was purely destructive. They were sure that their work was of the first importance to the human race. And what that work pointed to was, obviously, more and better rocketry. The advance of science seemed identified with the conquest of space, and both with the central interests of mankind.

Dyson, as a young physicist joining this circle, was plainly carried away by this tide. He still is. His case, however is complicated by a more adult conflict of motives than anything afflicting most of his American colleagues. He is a naturally humane and reflective man, inclining – as many theoretical physicists do – to a veneration for the whole universe, which leads him towards a kind of pantheism. His confident predictions of human conquest therefore alternate with expressions of awe and wonder. This makes his writings a good deal more impressive than those of the cruder exploiters, or of people who, though subtle, are prepared to be more openly brutal, such as Bernal and Haldane. If Dyson had managed to bring the two intellectual strands together, his contribution would have been invaluable. As it is, however, his increased authority only deepens the confusion.

DYSON AND MONOD ON MATTER

Dyson rejects Monod's fatalistic drama, not because it is depressing, but because it is unscientific. He rejects what he calls the 'taboo' which Monod and others have laid down against 'mixing science and religion'. This taboo, says Dyson, expresses an over-narrow view of what science is, a view which is out-of-date today, and moreover was only invented in the late nineteenth century. As an example of how earlier enquirers benefited by their freedom from it, he quotes a very interesting passage from Thomas Wright, the discoverer of galaxies, who wrote in 1750:

> Since as the Creation is, so is the Creator also magnified, we may conclude in consequence of an infinity, and an infinite all-active power, that as the visible creation is supposed to be full of sidereal systems and planetary worlds, so . . . the endless immensity is an unlimited plenum of creations not unlike the known universe.[1]

Wright, who was thus able to use a theological consideration to reach a correct conclusion in physics, added that, in so vast a system, 'the catastrophy of a world such as ours, or even the total dissolution of a system of worlds, may possibly be no more to the great Author of Nature, than the most common accident of life with us'. He explained that he found this a very 'chearful' idea, since it sets our own troubles in their proper proportion.

Dyson does not comment on this cheerfulness, though it is the opposite of his own reaction. What interests him is that Wright's Christian faith not only did not hamper his reasoning, but actually helped him to reach what are now accepted as sound scientific conclusions. Wright was certainly not alone here. Both Faraday and Clerk Maxwell were exceptionally devout men, active members of strongly Protestant Churches. Faraday, a Sandemanian, did not discuss his beliefs in scientific contexts, but Maxwell made it clear that his religion had been a great help to him in forming his theories. (Would the notion of Maxwell's Demon have occurred to somebody with a different upbringing?) The original forging of the modern understanding of electricity owed nothing to atheism.

Present-day commentators tend to be more distressed by this history than they might be if the great men had displayed some straightforward vice such as drunkenness. In his life of Maxwell, Ivan Tolstoy comments:

> Scientists who have attempted short biographies or essays on Maxwell have largely omitted the subject of his religion – passing over it, one assumes, in embarrassed silence.

We should remember, however, that the other two greatest mathematical physicists of our culture – Newton and Einstein – were also moved by a religious or at least a mystical spirit. To think deeply about the universe leads inevitably to an awareness of great mysteries . . . [Maxwell] reminds one of Einstein who was not only deeply concerned with epistemological questions, but was also something of a mystic, often bringing God into his arguments. 'God does not play dice' said he, when attacking the philosophic base of quantum mechanics, or 'God is subtle, but he is not malicious' and, 'I want to know how God created this world. I am not interested in this or that phenomenon, in the spectrum of this or that element. I want to know his thoughts; the rest are details.'[2]

Dyson thinks it irrational of recent practice to veto Wright's kind of thinking. The veto rests, he says, on an outdated idea of matter and does not

> take into account the subtleties and ambiguities of twentieth-century physics . . . The taboo against mixing knowledge with values arose during the nineteenth century, out of the great battle between the evolutionary biologists led by Thomas Huxley and the churchmen led by Bishop Wilberforce . . . A hundred years later, Monod and Weinberg were still fighting the ghost of Bishop Wilberforce . . . *In the bitterness of their victory over their clerical opponents, [the biologists] have made the meaninglessness of the universe into a new dogma.*[3] (Emphasis mine)

Dyson's talk of 'mixing science with religion' is vague. It is certainly better to speak about relating them. But then the taboo itself has always been vague. What is useful is that Dyson firmly ignores the emotive language which has so far served to protect this vagueness. He observes:

> Jacques Monod has a word for people who think as I do and for whom he reserves his deepest scorn. He calls us 'animists', believers in spirits. 'Animism', he says, 'established a covenant between nature and man, a profound alliance outside of which seems to stretch only terrifying solitude. Must we break this covenant because

the postulate of objectivity requires it?' Monod answers yes; 'The ancient covenant is in pieces; man knows at last that he is alone in the universe's unfeeling immensity, out of which he emerged only by chance.' *I answer no. I believe in the covenant. It is true that we emerged in the universe by chance, but the idea of chance is itself only a cover for our ignorance. I do not feel like an alien in this universe.* The more I examine the universe and the details of its architecture, the more evidence I find that the universe in some sense must have known that we were coming.[4] (Emphasis mine)

There are three quite distinct challenges to Monod here. Dyson is saying that:

1 mind, soul or life is native to the physical universe, not alien to it;
2 soul in some sense pervades that universe; and
3 *human* souls are in some way supreme in that universe and are destined to take charge of it.

Of these suggestions, (1) strikes me as plainly true and important. (2), which is sometimes called 'panpsychism', is an attractive and familiar but somewhat mysterious metaphysical notion; it needs much clearer explanation. As for (3) – the claim to human supremacy in the universe – that is surely quite separate from both the others and would need quite different support. It never gets it.

Dyson's conceptual tool-kit is, unfortunately, not well suited for distinguishing between these points. The only organized ideas he uses come from a few outlying areas of contemporary physics, brought in spasmodically at times to supplement personal avowals. It does not occur to him to look at the history of the dispute – at the way in which the different errors he attacks have arisen, or the kinds of alternative to them that have already been devised and distinguished. He shows no interest, for example, in the background of careful thinking that Christian scientists like Thomas Wright and Clerk Maxwell used and explained to justify their views. He proceeds as if he were confronting a hitherto undiscovered problem, and this makes it hard for him to move away from the way of thought he is attacking. His real and serious insights appear in striking flashes through the holes that he makes in that shared background. But they cannot be put together properly, because the only concepts he uses for stitching them come from other parts of physics itself.

ANIMISM AND MATURITY

Before looking further into this drama, it is worth noticing why Monod's taunts have been so effective in scaring off theorists more timid than Dyson. The notion of 'primitive animism'[5] comes from a familiar Enlightenment myth which compares the intellectual development of the human race to that of an individual. That myth gave the name 'animism' to a supposedly childish 'primitive' phase, followed, first by more organized religions, then by metaphysics, and finally, in the adult state, by science, which made all its forebears obsolete. Smaller, more 'primitive' cultures were always more childish than larger ones, and non-Western cultures, similarly, were more childish than that of the West. Finally, all other Westerners were more childish than Western scientists, who emerged as the only truly adult members of the species.

This claim to greater maturity has been important in forming the myths we are considering. Bernal, as we have seen, assumed that all other human occupations were childish or 'Melanesian', mere preparations for the life of science, which would soon supersede them. In the future,

Man will have anything from sixty to a hundred and twenty years of larval, unspecialized existence – surely enough to satisfy the advocates of the natural life. In this time he need not be cursed by the age of science and mechanism, but can occupy his time (without the conscience of wasting it) in dancing, poetry and love-making, and perhaps incidentally take part in reproductive activity. Then he will leave the body whose potentialities he should have sufficiently explored.[6]

Anthropology has not vindicated this kind of snobbery. All cultures are equally old, and people of different origins always tend to strike each other as childish, because they have received different kinds of social training. This impression of childishness, which is mutual, cannot support any value-judgments about relative maturity, either between cultures or between different groups in a single culture.

Within our own culture, we can clearly see that all occupations – including science – can be carried on either in more mature or more childish ways. There is certainly nothing specially immature or 'larval' about a life which is balanced rather than specialized. Indeed there is a well-known danger that people leading highly specialized lives will remain childish on whatever side of their personality is not involved in their work. And there is no obvious reason to treat doing science as the only mature aim.

If one insisted on making comparisons between cultures, there are obvious reasons why people in simpler cultures might count as *more* adult than highly civilized people, since they have to be much more self-reliant. By comparison with hunter-gatherers, most Westerners spend their lives cocooned in complex protective social webs, very much in the way that children do. This kind of childlike dependence may not be a bad thing, but it can hardly count as a sign of maturity.

OBJECTIVITY, DEADNESS AND ALIEN STATUS

Dyson has seen what is wrong with Monod's principle that 'nature is objective'. That principle means, as Monod insists, not just that we must accept things as they are, but that nature as a whole is somehow lifeless and inert, incurably an object. This lifelessness is what makes it alien to us and us to it, leaving us 'alone in the universe'.

Ignoring Monod's invitation to feel heroic, Dyson puts his foot through this piece of scenery. 'I do not', he says, 'feel like an alien in this universe.' Most of us will agree, but how should we explain the point? It might seem natural here to turn to biology and to talk about the relation between humans and other species. It would certainly be relevant to ask Monod what kind of entity he takes humans to be. In calling them aliens, he seems to be stealing Christian clothes, talking as if they were spirits from a non-earthly realm. What are these mysterious, purpose-owning beings, these minds which are alien to matter, but whose chief purpose is – for some strange reason – the study of matter, that is, scientific enquiry?

Dyson, however, ignores all such biological approaches and takes his stand on physics. He points out that Monod's notion of scientific method would not only 'exclude Thomas Wright from science altogether. It would also exclude some of the most lively areas of modern physics and cosmology.'[7]

THE DISSOLUTION OF HARDNESS

Dyson's sharp reaction against Monod's mean and short-sighted kind of reductionism is only a minor part of the wide protest it deserves. From a tactical point of view, however, it is extremely important. The special prestige of physics, its reputation as the root of all the sciences, makes this testimony crucial.

Physics is seen as the citadel, not just of intellectual thoroughness, but also of toughness, of hard-headedness. The odd metaphor by which physics and chemistry are called the 'hard' sciences has always had this implication of realism, of impartiality, of freedom from wish-fulfilment. This has, I think, looked plausible because of a scarcely noticed influence from the literal hardness of the physical subject-matter. The suggestion was that these studies, unlike all others, dealt with ultimate particles made of solid, rocklike, inert, impenetrable stuff. But this, of course, is an idea that physics itself has long abandoned.

Dyson explains well why other scientists still tend to cling to that imagery:

> It is easy to understand how some modern molecular biologists have come to accept a narrow definition of scientific knowledge. Their tremendous successes were achieved by reducing the complex behaviour of living creatures to the simpler behaviour of the molecules out of which the creatures are built. Their whole field of science is based on the reduction of the complex to the simple, reduction of the apparently purposeful movements of an organism to purely mechanical movements of its constituent parts. Every student of biology learns his trade by playing with models built of plastic balls and pegs. They are, for practical purposes, a useful visualization . . . But, from the point of view of a physicist, the models belong to the nineteenth century. Every physicist knows that atoms are not really little hard balls.

Every physicist, indeed, has known that since the early nineteenth century. But of course something more now goes with the loss of hardness. There is also a loss of determinism. Dyson goes on:

> For the biologists, every step down in size was a step towards increasingly simple and mechanical behaviour ... But twentieth-century physics has shown that further reductions in size have an opposite effect ... If, as physicists, we try to observe in the finest detail the behaviour of a single molecule, the meaning of the words 'chance' and 'mechanical' will depend upon the way we make our observations. The laws of subatomic physics cannot even be formulated without some reference to the observer. 'Chance' cannot be defined except as a measure of the observer's ignorance of the future. The laws leave a place for mind in the description of every molecule.[8]

Physicists, in fact, no longer use the mechanistic model of matter as inert, standard, homogeneous stuff. For their most interesting purposes, it no longer works, so they now see that it was never a literal description of the world. It was a metaphor based on analogies which proved fertile only for certain purposes. This change does not only affect matter itself, but also the idea of mind, because the two have been treated as correlates.

THE LURE OF MINDSTUFF AND MATTERSTUFF

This is a much wider point than Dyson's one about determinacy, and not dependent on it. Traditional talk of 'mind and matter' inevitably suggested that *mind* is the name of a distinct and more elusive kind of stuff or substance, something perhaps very thin and gaseous, but somehow parallel to matter. (Descartes himself clearly said that it wasn't, but his talk of 'two substances' made that idea very tempting). Since this gaseous stuff could not be found, there was a natural tendency to deny that minds existed at all.

To avoid that confusion, we need to move right away from this imagery of kinds of stuff. These words mind and matter are not the names of stuffs. They are much more like the names of viewpoints. If we talk about 'minds' we are attending to people (and animals) as subjects, as they are to themselves, from within. We are speaking of the way in which we and others experience the world. When we talk of 'matter' we are looking at the outsides of things, attending to items in the world around us as objects. If we then go on to talk about the relation between mind and matter, we will not be doing some kind of superstrange physics about the relations between unrelatable stuffs. We are talking about the relation between those two points of view, between subject and object.

That is talk about human life. It is philosophical psychology, and it needs to bring both viewpoints into play together. It cannot be carried on just in terms of physics, which is necessarily bound to the outward viewpoint, to considering things as objects. Psychology can indeed be 'objective' in the sense of being fair, unbiased and systematic. But it is still the study of subjects rather than objects, and it has to be carried on by methods suitable to that topic.

It is very lucky that physicists have now added their voices to those protesting against the crude, dualistic, mind-versus-matter schema, pointing out that the idea of 'stuff' is as ill-suited to the current physics of energy-fields as it always was to talk about minds. This mind–matter schema has again and again proved misleading when applied to complex phenomena, most obviously to those within human life. But so long as it seemed even faintly usable for physics, it still dominated the thought of people who were aiming to be scientific.

How far has it now retreated? As Dyson rightly says, current physics has now quite freed itself from the notion of hard, inert, billiard-ball-like atoms which used to be central to it, and molecular biologists are indeed lagging behind in still naively using that image. (How far they do lag can be noted in their constant exultant use of that spiral peg-and-ball construction to represent DNA.)

OTHER KINDS OF EXPLANATION

But the change has to involve much more than just this changed view of matter. It calls in question the special position of physics itself. The whole pattern of explanation that has for so long put physics at the centre of science is at stake along with the old view of matter. This change has scarcely registered yet in the public perception, including the perception of scientists. Physics is still described as 'the super-ego of the sciences', the model to whose condition they all aspire. And, in accordance with Descartes's old pattern, every hopeful suggestion about new elementary particles is still hailed in the press as offering 'the secret of the universe'.

But we ought to know by now that the universe has no single secret. It does not even have a single, central nest of secrets, to which some one study holds the key. We can explain a great many things, but in different ways. All studies are of strictly limited use; all are complementary, all need each other. They can be related, but not dragooned into unity.

Descartes's hope that physics might prove a glorious exception to this rule, the foundation on which all other studies depended, has been finally subverted by the development of physics itself. If there is no longer a fixed set of ultimate objects, nor a simple system of laws governing their movements, then there is no longer any reason to suppose that all explanation must terminate in the study of these objects' behaviour. The 'reduction' of larger to smaller phenomena, and the hierarchy of sciences which elevates the study of smaller ones always above that of the larger, has lost its justification.

USING PHYSICS AT ALL COSTS

Dyson and those who follow him have not seen this at all. Though he certainly wants a radical change in the way we think of matter, he does not at all expect this change to disturb the supremacy of physics. In his search for new ways of thinking to replace the old deterministic ones – a matter on which metaphysicians have made rather a lot of useful suggestions – he scarcely glances at any idea outside currently popular physical theories. Like other 'hard scientists' making this kind of move, he ignores most of human thought and looks for material for his new animism only in doctrines from his own speciality.

He does manage to find three places where mind can be held to impinge on physics, but they are terribly slight and patchy. His difficulties about them show plainly the crushing obstacle that exists to such a project. Modern physics has been deliberately devised to avoid all reference to subjects, and it has no tools for doing so. Trying to use it for this work is like digging with a sewing-machine.

The first meeting-point that he notes is sense-perception. Mind, says Dyson, meets matter when scientists look at their

data, a process which he describes in the following extraordinary way:

> At the highest level, the level of human consciousness, our
> minds are somehow directly aware of the complicated flow
> of electrical and chemical patterns in our brains.[9]

This is not just an inaccurate description, it is obviously false.
Sense-perception has been going on for many thousand years,
but, until quite recently, nobody at all was aware of these
patterns. Even today, most people are not aware of them, directly
or indirectly, nor are physicists themselves directly aware of the
patterns actually flowing in their own brains. What we *are* all
aware of in sense-perception is something quite different –
namely, the outside world, which we perceive through our
sense-organs. That perception, however, is not something that
can be described in the terms of physics. It would be business
for the biological and social sciences, and for common speech.

The next place where Dyson sees mind and matter as meeting
is quite remote from this one. 'At the lowest level, the level of
single atoms and electrons, the mind of an observer is again
involved in the description of events.' This is a reference
to awkward problems about the Copenhagen Interpretation
of Quantum Mechanics, something which we shall have to
consider because it is much relied on for these purposes.
Between this micro-level and sense-perception (Dyson continues), lies what is for these purposes a void, namely 'the level
of molecular biology, where mechanical models are adequate
and mind appears to be irrelevant'.

This is a quite extraordinary map of the intellectual scene. If
we are looking for cases where mind and matter are conceptually
involved together, we might surely expect some mention of the
fact that we can move our bodies. The reason why this escapes
notice is no doubt that Dyson's real hopes are pinned on the
Copenhagen Interpretation, and that interpretation calls only
for an observer. It does not mention that somebody is needed
to set up the experiment and move the glassware.

But of course the co-involvement of mind with matter is much
wider; it pervades every aspect of our life, and – to mention only
the learned – it has been dealt with in much detail by many
academic disciplines. Without these mediating areas where
the meaning of notions such as mind and matter might have

been more fully developed, Dyson is left treating these two categories still as stark alternatives. Though he wants to bring them together, he cannot help treating them as elements alien to each other. He is therefore forced to put his suggestion about their relations in a startling and dramatic form. He goes on:

> But I, as a physicist, cannot help suspecting that there is a logical connection between the two ways in which mind appears in my universe . . . I think our consciousness is not just a passive epiphenomenon carried along by the chemical events in our brains, but is an active agent *forcing* the molecular complexes to make choices between one quantum state and another. In other words, mind is already inherent in every electron, and the processes of human consciousness differ only in degree but not in kind from the processes of choice between quantum states which we call 'chance' when they are made by electrons. (Emphasis mine)

Why do the molecular complexes have to be 'forced' into action? Of course Dyson is right that consciousness is not just a helpless, passive reaction, driven from outside by alien 'chemical events', as Monod pictured it. Consciousness is an aspect of life, and it develops according to its own laws, which are a part of whatever laws life obeys. But this does not have to mean, either, that consciousness drives or 'forces' the physical processes, by interfering suddenly in the cracks between quantum states. There are not two rival kinds of stuff here, one of which must drive the other. There is a single coherent but exceedingly complex world, whose workings we understand only very partially in a number of different ways. The physical picture is as incomplete as all the others. It uses its own conventions and must not be mixed with others. It cannot be completed by grafting into it elements from quite a different kind of picture.

Dyson, however, having established this notion of a general, diffused consciousness pervading all matter, adds to these two 'levels on which mind manifests itself in the description of nature', a third, 'anthropic' level. He suggests that this conscious universe must have a purpose, that it is in some way designed to produce human thinking. Again, the reasons he gives for this belief are drawn only from physics. There are, he says, some 'striking examples in the laws of nuclear physics that seem to conspire to make the universe habitable', which it would not be unless there were some special reason for it. 'The peculiar harmony between the structure of the universe and the needs of life and intelligence is a third manifestation of the importance of mind in the scheme of things.'[10]

This third argument has grave problems of its own which we shall have to consider in chapter 17. But in any case, why consider only these three meeting-points? Once we have dropped the dogmatic assumption that life is meaningless, and the quite recently erected barrier dividing all thought about mind from thought about matter, we have clearly opened up again the whole wide field of previous human thinking on these topics. We can draw on a rich crop of ideas already formed about them, both in our own and other cultures.

We can follow these ideas through literature, anthropology and history. If we want to criticize or reject existing ideas, we can use the great conceptual tool-kit already forged for this work by moralists, metaphysicians, logicians, philosophical psychologists, anthropologists and historians, theologians, philosophers of science and philosophers of religion. And since people's deepest ideas about the meaning or meaninglessness of life are largely forged in everyday life and in the arts, we would surely do well to pay serious attention to these wherever we can find them. There is no customs regulation confining us to physics.

SHOULD WE SCRAP THE PAST?

Might all this body of existing thought be simply mistaken? Ought it all to be replaced by quite new ideas, drawn only from current physics? To read the books we are now considering you would surely think so, and Paul Davies has explicitly claimed as much:

> It may seem bizarre, but in my opinion science offers a surer path to God than religion ... science has actually advanced to the point where what were formerly religious questions can be seriously tackled.[11]

That is the advertisement; what is in the parcel? As his books show, Davies's claim depends on treating virtually all religious questions as depending on cosmological propositions centring on the Big Bang. But actually, not many questions of general importance do depend on views about that bang, however big. Only the most naive of fundamentalist theology treats the instantaneous creation of the world as philosophically important. Most religious questions arise within human life and begin by asking about its immediate meaning. The principles that we form in dealing with that area, which is real and immediate to us, rightly determine what kind of importance we shall attach to cosmological matters, if and when we come to learn about them. Our metaphysical ideas are rooted in the life that we know.

This is as true for physicists as it is for the rest of us. Any conclusions that specialists may draw about the relation of physical discoveries to life come from the whole of life, not just from physics, and are no stronger than their weakest link. Physics itself, moreover, is no self-contained enclave. Its arguments, like all other arguments, involve philosophical presuppositions, ideas that come from outside it. The questions involved in causal problems about the Big Bang are not internal to physics. They are shaped by crucial metaphysical notions about how causality, necessity, space, time, etc. should in general be conceived. Scientists who deal with these questions are doing metaphysics. They are perfectly entitled to do it and indeed must do it for these large, structural purposes. But whether their metaphysics leads them into religious thinking depends on all sorts of considerations internal to it and quite outside physical science itself. There is no short cut.

When we turn from causal questions to ones about purpose, Davies himself thinks we have gone outside the frontier of science. Indeed he rejects these aspects of the Anthropic Principle. After noting the apparent improbabilities of our existing world, he comments that 'many people of a religious persuasion will no doubt find support from these ideas' for their belief in providence, but 'those who prefer *a scientific perspective and language*' will deal with the matter by other means[12] (Emphasis mine). Davies, like Dyson and many others in this field, seems to recognize only two intellectual provinces. There is science – that is, physics – where everything is rational. There is also religion, which is licensed but wholly irrational, personal and inarticulate, a region where anything goes. The realm of physics is co-extensive with the realm of serious thought.

Now there has in fact been a great deal of serious pre-Dyson (and pre-Davies) thought about problems of mind and matter. No doubt it might be true that it was all mistaken. The temptation to assume so is naturally strong in academic areas where only the latest papers are read, and an article three years old is assumed to be useless. But in many central departments of human life this system of planned obsolescence does not work, because the essential problems are timeless. People who want to destroy vast tracts of existing thought in order to make conceptual revolutions therefore need to argue. They cannot simply proclaim that what they have not looked at is out of fashion.

What would be the effect of drawing all one's material for this kind of vast enquiry from current scientific theories? It would certainly not allow anything very lasting to be built. These theories exist in a constant ferment of change and development. Further large discoveries in theoretical physics are not just likely; they are positively expected and demanded. Anyone who based their metaphysical, and religious position – their whole attitude to life – on a particular interpretation of quantum mechanics, or on what has just been discovered at CERN, would be likely to need a new one every two years.

This transience is not just the mere general, unavoidable dependence we all have on the thought of our own age. That is a dependence on a great range of current attitudes. We gradually make slight changes in our own attitudes all the time, and we help to change the spirit of our times by doing so. The wide range of choices available within any age makes this general dependence on one's culture tolerable. But to tie one's whole personal attitude to the current findings of a particular science would be a far narrower commitment. It would be a prison indeed.

To build the understanding of something so immediately present and so central to us as our own nature on these highly technical, highly insecure theories is a desperate policy. If – as Dyson seems to suggest – these recent theories were the only ground for returning to the views that Monod classes as 'animism', they would not be an adequate one. In that case, nobody up to the present age would have had any good reason to accept those views, and if the fashion in physics changed again, people in the future would have no good reason to go on accepting them either.

Historically, however, these 'animistic' ideas flow from a long, rich and well-considered tradition. The whole range of subtle language which enables us to discuss them at all is something much larger and older than any modern physical theories. It expresses much perfectly sensible thinking on the matter done by generations of people who did not know, and did not need to know, that such a study as modern physics would ever exist at all. Dyson is quite right that modern physics has removed a certain block which earlier physics had put in the way of such thinking. But that does not mean that it can provide, on its own, the way forward.

SCIENCE AND NON-SCIENCE

Dyson, however, does see his animist reasoning as being still inside the province of science, though only just. 'This', he says,

> is as far as we can go as scientists. We have evidence that mind is important on three levels. We have no evidence for any deeper unifying hypothesis that would tie these three levels together. As individuals, some of us may be willing to go further. Some of us may be willing to entertain the hypothesis that there exists a universal mind or world soul which underlies the manifestations of mind that we observe. If we take this hypothesis seriously, we are, according to Monod's definition, animists. The existence of a world soul is a question that belongs to religion and not to science.[13]

Here, again, is that very odd map of the intellectual world, a map not peculiar to him, but shared by most of the people who concern us. It shows only two areas, science and subjectivity. Outside science – which, as we have seen, means 'outside physics' – everything is irrational.

No province of thought is marked on this outer region of the map except religion. And religion itself is viewed as something subjective, non-rational, privatized, something we can only decide about 'as individuals'. It is not suspected that there might be better and worse ways of thinking about it, more and less intelligible suggestions, concepts we can develop together by public discussion, standards of reasoning by which we can help each other. The names of two provinces of thought which might be thought relevant – theology and metaphysics – do indeed sometimes appear, but usually just as warnings, synonyms for senselessness – 'here be dragons'. All other kinds of methodical thinking are simply ignored.

This works very oddly. To accept the existence of a 'unifying mind or world soul' as Dyson suggests , is plainly a structural proposal, a move liable to bring together other concepts that do not make sense without it. Dyson, however, speaks of it as an extra hypothesis, a belief which we may personally 'entertain', but for which there is 'no evidence'. That makes it sound like an empirical guess about, for instance, an extra planet, a story that ought to have observations to confirm it, but has not yet got them. But that cannot possibly be its proper standing.

16

SPACE, FREEDOM AND ROMANCE

COLONIZING ZEAL

So far, Dyson has been helpful. He rightly refuses to be bullied into isolating science from its conceptual background. He sees that some notion of soul or mind is needed to make sense of the universe. He is right, too, to shout that he feels at home in the cosmos, and that we ought to take seriously the question of where we feel at home.

But that question is just where the difficulties start. Dyson is a dedicated and intensely romantic devotee of space-colonization, which he treats as inevitable, and hails as the prelude to the mechanical changing of humanity into deathless, non-organic forms. This commitment raises questions which are not just questions about whether somebody, some day, will explore space. That is a factual issue which may never arise, and which cannot concern us now. They are questions about the reasons for desiring and promoting this project today, about its expected function, about what the proposal is supposed to do for us.

The reasons Dyson gives for his enthusiasm are mostly familiar. Where he is original is in claiming to literalize them, to turn his quasi-religious faith in the immortality and supremacy of the human intellect into a regular branch of physics. Bernal and Haldane had published these schemes only in popular books. Dyson does this too, but he also puts them in technical form in learned journals, in articles consisting mainly of equations. Yet in doing this he keeps the exalted tone, the eager evangelical spirit, and the explicit offer to restore meaning to life which marked his Marxist predecessors. Here is salvation through science indeed.

As we have seen, at least some other respected physicists accept his claim. Notably, Barrow and Tipler welcome Dyson as the founder of 'the new study of "physical eschatology"', 'the study of the survival and the behaviour of life in the far future', a study which has, they hold, become legitimate now that discussion of it is 'based entirely on the laws of physics and computer theory, in sharp contrast to the vague speculations which were typical eschatological discussions prior to Dyson'.[1]

It might seem reasonable to ask how any study of 'the survival and behaviour of life' at any time can possibly be considered as part of physics, since 'life' is not a physical term. In the parallel case of speculations about possible life-forms elsewhere in the universe, the name 'exobiology' has been coined. Whatever the value of such enquiries, this name does at least pay lip-service to the existence of the life-sciences and to their difference from physics. Here, by contrast, physics appears, without any apology, as the only possible alternative to 'vague speculations'. It is accompanied, however, this time by computer theory, a branch of thought which has, for depressingly obvious reasons, now managed to break the tabu ruling all other ways of thinking out of court.

To return however, to Dyson – in his autobiography he quotes from his 'Space Traveller's Manifesto' written in 1958, when he was working on a project for the US government to build spaceships powered by small atomic bombs . . . He had, he says, been convinced from his childhood that men would reach the planets in his lifetime and that he would help in the enterprise. This conviction rested on two beliefs, one scientific and one political:

1 There are more things in heaven and earth than are dreamed of in our present-day science. And we shall only find out what they are if we go and look for them.
2 It is in the long run essential to the growth of any new and high civilization that small groups of people can escape from their neighbours and from their governments, to go and live as they please in the wilderness. A truly isolated, small and creative society will never again be possible on this planet.[2]

In 1958, such talk was of course common, but Dyson does not criticize it in 1979. Discussing then the various possible ways of travel, he declares a preference for small, light, cheap spacecraft, and asks plaintively,

> When will the third romantic age of the history of space-flight begin? The third romantic age will see little model sailboats spreading their wings to the sun in space, as free and graceful as the little radio-controlled gliders which dance among the birds in the sea-breeze over the cliffs near the General Atomic Laboratories every Sunday afternoon. It will see test-stands as amateurish as those of Berlin and Point Loma, where a new generation of young people will try out a new generation of wild ideas.

SPAM IN CANS

What sense does this make today? Quite apart from the fact that we are living in an ecological crisis far too grave to make financing such schemes realistic at all, we have, since 1958, seen the Apollo project. This dose of reality should surely have produced doubts about both his objectives. For scientific purposes, manned spacecraft have proved far less useful than unmanned probes. When human beings are sent up, nearly all the effort of the engineers has to be expended on keeping them safe, healthy and likely to return alive, so that research becomes a mere side-issue. It is also, of course, bizarre to identify the whole of 'science' with astrophysics when we still have an infinite amount to discover about our own planet.

Dyson's second aim – personal freedom – is even less compatible with real space-travel. What goes up is not just a set of minds; it is bodies with their usual drawbacks. For astronauts so far, space travel has been heavy stuff, imposing a more rigid discipline than any religious order – a discipline that no one except a few invalids endures in normal life.

Though spacemen struggle, as Tom Wolfe reports, against the sense that they are essentially 'spam in a can', and though they have managed, over time, to gain a little more control over their vehicles, they still pay the price of total conformity on all personal matters. Every action, every mouthful of food, every sleep and every defecation has to be as their directors at the computer terminals on earth dictate. They are, too, more closely crowded, more pressed against each other, than anyone would normally find bearable. They do not 'live as they please in the wilderness'. Indeed space, in the ordinary sense, is a luxury quite unknown to them.

Of course this is supposed to be only an early stage in space-travel's development, a stage which Dyson rather dislikes. He thinks that 'huge and politically oriented programs like Apollo are perhaps not even going in the right direction . . . The road that will take mankind to the stars is a lonelier road.'[3]

But it can no more be lonely than mining can. Where there is no air to breathe, no ground to tread on and no natural protection from heat and cold, animals with bodies like human beings must always be totally dependent on what is done for them by others. Their incredibly expensive equipment has to be corporately funded. They cannot possibly have individual freedom. Like engine drivers, they have some power, but only as agents of a huge, impersonal group.

POLITICAL WORRIES

Remarkably, Dyson admits all this. But it only makes his dream retreat, as dreams tend to do, to a hazier, more distant scene. These troubles are, he says, merely a limitation of the inner solar system. Emigrants there will indeed have to make do with 'gray technology', building

> colonies in space in the style of O'Neill's 'Island One', cans of metal and glass in which people live hygienic and protected lives, insulated from both the wildness of earth and the wildness of space. We will be lucky if the people in these metal-and-glass cans do not come to resemble more and more as time goes on the people of Huxley's *Brave New World*.[4]

But Dyson is not worried by this. *Brave New World* is not to be more than a passing phase. He is not upset, either, by the political problems of sacrificing the whole present and the near future to remoter possibilities. Apparently, he expects the political will that existed for a short time as a by-product of the cold war to continue indefinitely and to be multiplied so as to provide for these sacrifices.

He must also be assuming – rather surprisingly – that people will not become so stunted or corrupted by these early stages as to make further developments either impossible or useless.

If the population of the inner solar system were indeed reduced morally to a *Brave New World* condition, how could it be expected to generate the ideal communities he is planning? Schemes like this always assume, too, that the scientists themselves will never become stunted or corrupted, and that the populace will always leave them in full charge of cosmic events. None of this seems very realistic.

GREEN VARNISH

To continue the dream – later on, Dyson explains, things will be very different:

> Green technology pushes us in the right direction, outward from the sun, to the asteroids and the giant planets and beyond, where space is limitless and the frontier forever open. Green technology means that we do not live in cans but adapt our plants and animals and ourselves to live wild in the universe as we find it. The Mongolian nomads developed a tough skin and a slit-shaped eye to withstand the cold winds of Asia. If some of our grandchildren are born with an even tougher skin and an even narrower eye, they may walk bare-faced in the winds of Mars.[5]

Must we suspect that an eminent physicist either has no grasp of biology or is not particular about telling the truth? Dyson leaves us little choice, especially when he goes on to say that this process is just part of the normal

> expansion of all life, making use of man's brain for her own purposes . . . Our spread through the galaxy will follow her ancient pattern.
>
> To make a tree grow on an asteroid in airless space by the light of a distant sun, we need to redesign the skin of its leaves. In every organism the skin is the crucial part which must be delicately tailored to the demands of the environment.[6]

This is simple madness. The biological term 'adaptation' cannot be used to describe what might happen if somebody charges in and decides to redesign an organism. In biology, 'adaptation' means something which organisms do of their own accord and within their own repertory.

Dyson's technology is not 'green' except in the sense that it is directed at green subject-matter. He can only represent this violent intervention as part of the organic process by personifying 'life' as a deity, an independent agent who is on his side. 'Life' is arbitrarily given the purpose of completely transforming trees in this way – a purpose which luckily coincides with his own and that of like-minded humans. This personifying of 'life' and hijacking of it for one's own fancies is not a fact of biology, nor a conclusion of physics. It is Bernard Shaw's tendentious myth of the Life Force – a myth which may have its uses, but which carries no scientific authority and incurs serious moral costs.

THE FRAGILITY OF ORGANISMS

The myth works here to obscure the central fact about adaptations, namely, that the scope for them is very limited and most of them fail. Organisms are radically dependent on their habitual environment, not just in a few obvious ways, but in an infinite number of less obvious ones too. (If you 'redesign the skin', what about the lungs or the eyes or the stems or the roots?) Quite slight changes, even within ordinary earthly conditions, are constantly destroying species and, when species go, habitats can go with them.

This fragility has become frighteningly plain with the loss of soil-fertility following the destruction of tropical forests. Great parts of the earth's surface which could in principle be fertile, and once were so, have become lifeless because of some disturbance, often because of human action. But the asteroids (and comets) which Dyson is recommending have no soil, no atmosphere, no gravitation sufficient to hold a soil or an atmosphere if these were provided. The giant planets have gravitation so excessive as to crush most earthly structures; their atmospheres are turbulent and fearfully unsuited to life. Both kinds of body have surface temperatures far colder than even the simplest earthly life-forms can tolerate. What sense does it make to suppose that 'life' – not just in bacterial forms but in communities that could support human colonists – could be transferred to them at all? Yet Dyson predicts:

> Millions of little worlds, conveniently accessible from earth, where suitably programmed trees could take root and grow in the soil as they find it. With the trees will come other plants, and animals, and humans, whole ecologies in

endless variety, each little world free to experiment and diversify as it sees fit.[7]

ALCOHOL-TREES AND LIVING PIPELINES

In his green mood, Dyson commends these schemes as defences against our environmental crisis – as retreats made necessary by the destruction of our own planet. This may strike some of us as rather like grabbing an umbrella and throwing yourself off the roof because the house needs repairing. Why should anyone think these implausible methods more promising than a direct attempt to save the earth that we now have?

What makes the implausible methods seem plausible is, it seems, the magic of DNA. Dyson has already described uses for this magic on earth:

> Imagine a solar energy system based on green technology, after we have learned to read and write the language of DNA so that we can reprogram the growth and metabolism of a tree. All that is visible above ground is a valley filled with redwood trees, as quiet and shady as the Muir Woods below Mount Tamalpais in California. These trees do not grow as fast as natural redwoods. Instead of mainly synthesizing cellulose, their cells make pure alcohol or octane or whatever other chemical we find convenient. While their sap rises through one set of vessels, the fuel they synthesize flows downwards through another set of vessels in their roots. Underground, the roots form a living network of pipelines transporting fuel down the valley. The living pipelines connect at widely separated points to a nonliving pipeline that takes the fuel out of the valley to wherever it is needed. When we have mastered the technology of reprogramming trees, we shall be able to grow such plantations wherever there is land that can support natural forests . . . If we assume that the conversion of sunlight to chemical fuel has an overall efficiency of one-half per cent, comparable with the efficiency of growth in natural forests, then the entire present energy consumption of the world could be supplied by growing fuel plantations on about ten per cent of the land area.[8]

The use of apparently exact figures in this passage is depressing.

They give an air of responsible background research, but they are entirely meaningless. It is surely startling to see how quickly, in an unfamiliar area, highly-qualified scientists can revert like this to purely magical thinking. Dyson and other similar sages assume that genetic engineering can do absolutely *anything*, and apparently at no cost – the free lunch at last. Just so, a few decades back, the same thing was supposed about nuclear energy. In both cases, the actual successes at the time when the faith took root have been very slight, and these successes have, of course, been confined to fields where the prospects looked peculiarly promising.

It seems to be forgotten both that science and technology have had endless failures, and that their successes have always depended on concentrating on the things that work and avoiding what does not. In designing 'futures', a few mildly hopeful existing omens are extrapolated without limit. With hardly anything presently in the bank, blank cheques are drawn on behalf of chosen techniques – a process which continues long after their limitations have become obvious.

If, however, miracles like these could really be worked on earth, why would there be any need to think about using space at all? Compared with the problems of colonizing an asteroid, or even Mars, the work of restoring plentiful vegetation all over the earth, including the Sahara and the Antarctic, and of designing a problem-free contraceptive that would enable stable human populations to live there, would be child's play. There could be no need for emigration. Certainly all this would presuppose peace, good government and co-operation. But then, the project of space-colonization must presuppose that anyway.

WHAT FUTURES ARE

All the practical reasons that Dyson and others cite for launching out into space – garbage disposal, mining, pollution-free industry, increase of population and the like – depend on economic claims which used once to look plausible because they were so extremely distant. Dyson still cheerfully makes them; for instance, 'the resources of this planet are finite, and we shall not forgo for ever *the abundance of solar energy and minerals and living space that are spread out all around us*'[9] (Emphasis mine). Again, this is not meant to be fiction. Since the moonshots, space projects have left the world of Gulliver, where wildness could be a virtue, and arrived in that of public accounting, where it is not.

Why is this so hard to grasp? Perhaps we should attend a little to some problems about our current idea of 'the future'. As first envisaged by Nietzsche, Wells, the Italian Futurists and its other earlier prophets, the Future was not exactly a name for what could be expected to happen. It also functioned as a blank, a clear field for fantasy, a realm where desired situations could be projected. It took the place of distant countries and Golden Ages set in the past, both of which were becoming unconvincing in a better-informed age. 'The Future' was, of course, different from both these in that to talk of it always involved extrapolating certain trends from the present. But these trends could be chosen at will according to the writers' interests. Inconvenient factors could just be ignored.

In a fast-changing world, however, these prophecies inevitably began to include also an increasing element of literal prediction, and of proposals seriously meant to be adopted. They also began, following Wells, to concentrate more and more on the hardware of technology. Nietzsche, who always dealt with psychological and spiritual issues, would have seen how misleading this was. He would have insisted on directing attention to the symbolism of machine-worship itself. He would have hunted out and separated the various motives underlying these fantasies. It is a great pity that he did not survive to do this bit of analysis and traffic-direction.

As it was, confusion continued. Real hardware was rapidly changing, and often did, of course, actually embody the dreams of its designers. Fantasy therefore remained hopelessly entangled with practical proposals for research and development. The discussion of possible inventions could be freely used for wish-fulfilment, expressing dreams that are at best religious and at worst just mindlessly greedy and destructive. The bloody fancies of the Italian Futurists, who had no understanding of machinery but revelled in its anti-human symbolism, did not stand alone. The whole intense development of weapons, and of much other psychologically seductive hardware at the expense of more necessary objects, has plainly owed much more to dreams of domination than to sober calculation of likely benefits for anyone.

There is also a curious moral slant, which tends to condemn all opposition to the wilder technological dreams automatically as narrow and philistine. The response felt to be 'scientific' is not the shrewdly sceptical one which at one time was seen as typical of science. It is a receptive, credulous, romantic one. The fancies themselves are venerated as being, not just wish-fulfilment, but embodiments of a serious ideal. They could, of course, in principle be so. But if they are, then the nature of that ideal needs to be spelt out soberly and literally. It cannot be taken on trust in a flood of rhetoric.

FREEDOM AND DIVERSITY

People who want faiths will build them somehow. Because, earlier in the Industrial Revolution, things often proved possible which had not seemed to be so, it is easy to feel that, given enough science and enough confidence, possibilities now really are infinite. These possibilities extend, of course, to successfully redesigning ourselves. Accordingly, after describing the spiritual confinement of those living on the inner planets, Dyson goes on:

> Humanity requires a larger and freer habitat. We do not live by bread alone. The fundamental problem of man's future is not economic but spiritual, the problem of diversity. How do we find room for diversity, either on our crowded earth or in the metal-and-glass cans that our existing space-technology provides as living space?

'We' must (he argues) therefore provide both for social diversity through custom, and also for

> diversity on the biological level [which] means allowing parents the right to use the technology of genetic manipulation to raise children healthier or longer-lived or more gifted than themselves. The consequence of allowing to parents freedom of genetic diversification would probably be the splitting of mankind into a clade of non-interbreeding species.[10]

This extraordinary diversification of humans was, as we have seen, proposed by Haldane and Bernal and accepted by Day; it seems to be now a regular part of these future-myths. Why anybody should expect or want it is mysterious. Prospective parents, if really given these powers, would surely be more likely to produce an unusually homogeneous generation of children, designed according to the fashions of their time and culture. Genetic engineers, sharing the parents' background, would probably help them to do so. No doubt they would advise some diversity, but they would only see the possibilities that attracted their own generation. It is still less clear why either party would ever want the next generation to be divided into unmixable tribes.

Much odder than that, however, is the general assumption that more diversity is so badly needed. Why do parents suddenly need the right to make their children more varied? The idea is strange because, as things are, a great part of education in every culture is directed towards making people *less* varied than they naturally are – towards ironing out their differences. These efforts often fail, leaving a lot of people lamenting, as Blake did:

> Oh why was I born with a different face?
> Oh why am I not like the rest of my race?

If we actually want more diversity, we need only relax these educational efforts to get it at once. Do we really want it? Or do we just want something better? Diversity beyond a certain point within a species does not seem to be a great good in itself. It can easily become a serious evil by blocking co-operation. Dyson indeed points out this difficulty, but he turns it into one more argument for space-colonization, since socially incompatible sub-species must be kept apart.

SPACE, LITERAL AND METAPHYSICAL

The interesting thing about all this space-planning is of course the spiritual reasons given, the ideas expressed about non-physical needs which these schemes are to satisfy. Symbolically, space stands for freedom. Negatively, this means not being interfered with by others; positively, it means increased opportunities for action. Dyson no doubt has both in mind in his talk of diversity, and again when he says that we have a

spiritual need for *an open frontier*. The ultimate purpose of

space travel is to bring to humanity, not only scientific discoveries and an occasional spectacular show on television,
but a real *expansion of our spirit*.[11] (Emphases mine)

These are metaphors. Real space, in the sense of the area outside
the earth's atmosphere, is indeed physically larger, but seems
quite unsuited to provide what the metaphors stand for. It has,
of course, been drawn into this kind of metaphorical use as heir
to the literal, geographical 'frontier' which Americans were for
a time accustomed to see as symbolizing unlimited freedom,
always available to them if they chose to use it. It was usually
forgotten that what lay beyond this frontier was not actually
empty land or a new kingdom of the spirit, but territory already
inhabited by other people and animals. Considering the history
of colonization, it is surely remarkable to find this imagery being
used with so little embarrassment today.

There was thus a chronic confusion between the possibility of
inner, spiritual freedom and the possession of an outside, physical territory which one could, if one felt like it, always invade.
Eastern cultures, where people have long lacked such fields for
physical expansion, do not seem to make this confusion. They
have given much more attention to the direct cultivation of
inner freedom. This is no doubt one reason why they are now
becoming so popular in the West.

Of course that project has it limits. External crowding as seen
on the Tokyo underground is excessive by anyone's standards,
and the Sony Walkman has been deliberately devised for
self-protection in such circumstances. Human over-population
is indeed a terrible menace, one which genuinely does require
help from technology – notably, of course, through contracep-
tion. But technology can at best only supplement our primary
inner methods of tolerating other people and finding our own
fulfilment. Dyson wants to provide for both aims by universal
access to space. ('Space travel can only benefit the mass of
mankind if it is cheap and generally available. We have a long
way to go.')[12] But why this should be the right direction to go
in remains quite obscure.

17

THE ANTHROPIC SYNTHESIS

BRINGING IT ALL TOGETHER

We come now to the book which deliberately conflates the various prophecies mentioned so far, overlooking the vast differences in the spirit and intention behind them, updates them, fills in the gaps, adds some curlicues of its own, and backs the scientific status of the whole by a mass of detailed calculations. In one way, *The Anthropic Cosmological Principle* is a welcome book. It does open a lot of long-closed doors. Barrow and Tipler know that the large questions physics raises cannot be handled with tiny tools. They enjoy the vast perspectives. Like Dyson, they also like shocking people with a narrower view of science. And, having read much more widely than he has, they bring together a mass of fascinating material, some of which really does illuminate their topic.

However, though their materials are so wide, their choice of methods is not. Their vision – and it is a real vision – is to bring all these questions within the province of 'science' defined as they now define it, that is, to handle them all by the methods of current physics. We have already seen how pleased they are to claim that discussion of 'the survival and behaviour of life in the far future ... is now based entirely on the laws of physics and computer theory ... in sharp contrast to the vague speculations which were typical eschatological discussions prior to Dyson'.[1]

To bring such an ocean into this pint pot would certainly have been a remarkable feat. Their view of the scope of the problems is actually a good deal narrower than Dyson's. They do not follow up his ideas about animism and the nature of minds. Instead they settle briskly for some simple reduction of minds to information. The details of this reduction are none too clear, since they are happy to write, on a single page, that a human being 'is fundamentally a type of computer', that it is 'a program designed to run on particular hardware called a human body', and also that it is 'a representation of a definite program rather than the program itself'.[2] Since they are satisfied to define an intelligent being behaviouristically, as one that can pass the Turing Test, they evidently do not think consciousness is necessary for it, although, again in the same passage, they cheerfully explain that 'we might even identify the program which controls the body with the religious notion of a soul, for both are defined to be non-material entities which are the essence of a human personality'. All this casualness, however, makes the general message plain. Information theory now saves us the trouble of paying any serious attention at all to the nature of thinking subjects.

Things are different, however, when it comes to teleology. Purposive reasoning is something they really do need. They are committed to 'a progressive Cosmos, evolving towards a higher state',[3] and they discuss with some care and respect the work of philosophers such as Hegel, Schelling, Bergson, Alexander and Whitehead who have argued for such a cosmos. Indeed they make a highly laudable search for intellectual ancestors, showing how many thoughtful scientists, as well as philosophers, have distinguished between various kinds of teleological thinking and used various forms of it effectively. This historical discussion can certainly be helpful for future work.

There is a wild gap, however, between these careful researches and the startling conclusions they reach about cosmic destiny. (As usual in these scientific fantasies, Jekyll and Hyde seem to seize the word-processor by turns.) In their scrupulous mood, they prepare for this jump by explaining that what they mean by teleology is not found in what would seem its most obvious place, namely living organisms, but in cosmology:

> The simpler teleological arguments concerning biological systems were supplanted by Darwin's work, but the system of eutaxiological arguments regarding coincidences in the astronomical make-up of the Universe and in the fortuitous

forms of the laws of Nature were left unscathed by these arguments that have evolved into the modern Anthropic Principle.

What they mean by eutaxiological is never really clear, nor is the distinction consistently observed. However, the kind of thing they finally approve presumably explains the point. For instance:

> Were one to adopt a teleological view of Nature, one could go so far as to assert that matter has many of its properties today not because these properties are necessary for life today, but because these properties will be *essential* for the existence of life in the distant future. However, we would expect such teleological properties to exist in matter only if SAP were true, and that life is in some way equally essential to the Cosmos. Are there any reasons to think that life is essential to the Cosmos?[4]

They give two reasons. One is the proposal already mentioned for prolonging the life of the universe by dumping matter down black holes, which they stress is 'extremely speculative'. The other, more central to them, is the view about the necessity of observers, arising from what they admit are 'rather controversial interpretations of quantum mechanics'.

It might seem at this point that they are distancing themselves from both these arguments as too 'teleological' rather than 'eutaxiological'. But this would not be right. In discussing the argument about observers earlier, they say, with much greater confidence:

> In this chapter we have seen how modern quantum physics gives the observer a status that differs radically from the passive role endowed by classical physics. The various interpretations of quantum mechanical measurement . . . reveal a quite distinct Anthropic perspective from the quasi-teleological forms involving the enumeration of coincidences which we described in detail in the preceding two chapters.[5]

This claim to a standing independent of teleology seems, however, to be wish-fulfilment. The perspective would not be anthropic if it were not already teleological – if the universe did not demand these observers, and produce them because it needed them.

THE ROLE OF THE COSMIC COINCIDENCES

It is time to summarize the anthropic argument itself.

Officially, it begins from some considerations, widely accepted among physicists, about the apparent improbability of the existing universe. The basic construction of the physical world appears to rest upon some surprising coincidences. The conditions that make possible the existence of our bodies and of all the things we know are not, it seems, necessary or even probable. They depend on a set of 'fundamental constants of Nature; for example, quantities like the electric charge of the electron, the rate of the electron and proton masses'[6] which seem surprisingly well adapted to fit together.

All these might, as far as is known, have been quite different, and, had they been so, a different world (if any) would have resulted. Among the different worlds that there might possibly have been, it seems that very few could possibly have harboured life at all, still less intelligent life. 'The subset of cognizable universes, amongst a collection in which the constants of Nature take on all possible permutations of all possible values, is very small.' We are, in fact, extraordinarily lucky to be here.

Physicists are naturally much interested in investigating the details of these fundamental constants. This approach, say Barrow and Tipler,

> typifies the modern reaction to the facts that fuelled the Design Arguments of past centuries; but whereas the ancients might regard it as a mark of divine favour that the earth possesses a life-supporting atmosphere while the moon does not, now it would be more immediately attributable to the fact that only bodies exceeding a certain critical size will exert sufficient gravitational pull to prevent gas molecules escaping.

Natural explanations must, in fact, be followed as far as they can go. Yet at present they stop at the Fundamental Constants, which is unsatisfactory. Enquiry does not normally resign itself at such points. Of course we might really have reached an impassable terminus. But we ought not to assume that without good reason.

Orthodox science therefore continues to seek for further natural explanations of the accepted kind.

METAPHYSICAL EXPLANATIONS: MAN REPLACES GOD

Barrow and Tipler, however, shift gear sharply here to a quite different type of explanation, and this shift is what first calls for our attention. The 'anthropic' explanation they offer, though officially still scientific, is certainly metaphysical. It may either be classed as frankly teleological or may try to save itself from that awkward category by falling under the Many Worlds Interpretation of Quantum Mechanics – which is, however, itself also a very ambitious piece of metaphysics.

The proposal is that this world – among all possible worlds – is one that has been made real by being observed by human scientists. This surprising past history, however, is not all. Many central themes in the book – the insistence on space-colonization, the guarantee given by the Final Anthropic Principle and the prediction about the Omega Point, etc. – indicate that this remarkable process was only the first step in the world's destiny. Scientists, having made the world real, will also fulfil its ultimate purpose.

It is hard to see how this singular type of explanation could actually help with the scientific problem of explaining the cosmic coincidences. The use of it seems, in fact, to mirror the many, now discredited, arguments which invoked God to fill current gaps in scientific reasoning – a 'God of the gaps'. A notorious example was Descartes' positing of a perpetual miracle to keep soul and body united in despite of their radical incompatibility.

Scientists who used the notion of God in that way were not answering the scientific questions raised by strange facts about the world, but doing something quite different. Usually they were reassuring themselves and their readers that these surprising facts could still be contained within their world-picture; that the universe had not stopped making sense. For people who already believed in God, this was in itself a legitimate proceeding, a mere reminder that he was still in charge. It only became dangerous when it was used – as it sometimes was – to block further enquiry.

199

Yes. *How does*
this help *people*
who don't
believe in
God

For people who do not already believe in God, however, these arguments are vacuous. The frequent attempt to use them as a proof of God's existence – to claim that these mysterious facts, inexplicable without God, demand God as an explanation – failed, because he simply was not an explanation of the right kind at all for these factual puzzles. Moreover, as theologians have pointed out, too, the concept of such a 'God of the gaps' has no value from a religious or moral standpoint either. All that could possibly be deduced from one of these unsolved factual puzzles is that some unknown quantity – some x – must be there to resolve it. Unless you already have views about what that x is, this really means nothing. It certainly could not imply a good God, nor a God as conceived in any particular tradition.

YES!
Anti-view

Barrow and Tipler seem here to be invoking a Man of the Gaps. Believing devoutly already in the supremacy of Intellectual Man, they see him as what gives sense to the universe. They therefore invoke him as its kingpin with as much confidence, and as little suspicion that they have stepped outside science, as their predecessors felt about God. They see Man as shaping the universe in such a satisfactory way that it is only natural to treat the fundamental constants as part of his natural equipment, pointers to his importance, factors determined in some way by his needs.

REMOVING THE GAPS

We can see the irrelevance to science of this kind of explanation – whether it invokes God or man – by asking what happens if the scientists do manage, after all, to solve the puzzle that they now find insoluble. Supposing that next year's Nobel Prize winner comes up with a perfectly satisfactory explanation of the cosmic coincidences, will the Strong Anthropic Principle then have been disproved or made unnecessary? If not – if it still stands – can it really be deriving any strength from these coincidences now? (Barrow and Tipler mention this query as having worried an earlier theorist, and call it 'the bugbear of all Anthropic Principle arguments',[7] but they do not seem uneasy about its effect on their own.)

From Hume's or Monod's position, of course, the anthropicists' mistake is to believe that the world makes sense at all. This (as I have already suggested) is not a helpful or rational response.

200

If we mean to go on thinking about the world we have to think that in principle, somehow, at some level, it does make sense.

But very many views are possible about what makes it make sense – about what constitutes our salvation. The anthropic proposal is only one possible suggestion, and in many ways a most peculiar one. It is surely being put forward because its authors feel the need of teleology for exactly the same reason that other people do – because they are not satisfied with an unintelligible world – and are already so fully committed to this kind of faith that they see it as the only possible candidate. They disguise it heavily in scientific clothes because they think its naked form would be indecent.

WHY JUST US?

If this dismissal seems a trifle brusque, the next thing to notice is how weak the argument from improbability actually is. Why, for a start, should the cosmic coincidences point particularly to us, to human beings? We are not the only living things, nor are living things the only part of the universe that owes its existence to this special cosmic situation.

Any set of complex conditions that has been operating in isolation for a long time is bound to produce very many things which could not have existed in other conditions. Whatever the universe had initially been like, it would by now have contained a great mass of things that could exist only in it. Stephen Jay Gould puts this point sharply in discussing Alfred Wallace's version of the Strong Anthropic Principle:

> The central fallacy of this newly touted but historically moth-eaten argument lies in the nature of history itself. Any complex historical outcome – intelligent life on earth, for example – represents a summation of improbabilities and becomes therefore absurdly unlikely. But something has to happen, even if any particular something must stun us by its improbability.[8]

– a point which he has developed with great force at more length in *Wonderful Life*.[9]

The improbability of human existence, then, is not in itself at all improbable or rare, not a unique distinction. The universe must unavoidably contain an immense number of such improbable things. Endless non-living things, as well as all the living ones, share this situation with us. For instance, the impressive red spot that satellite-pictures reveal on the flank of the planet Jupiter may well, as far as I know, be something that could occur in no other universe than ours. If it is, this would probably not make anybody think that the universe was designed specially to produce that red spot.

What, then, about the mass of living things around us? It seems often to be assumed in these arguments that other animals and plants are markedly less improbable than ourselves – that our form of intelligent life is uniquely unlikely. This seems a very strange idea. By what standards could one weigh the improbability of our hypertrophied cerebral cortex against – say – the improbability of a giraffe's hypertrophied neck and legs, or of a fiddler crab's single gigantic claw? What, again, about the improbability of giant pandas, taxonomic carnivores who are complete vegetarians and live only on a single species of bamboo? What about colonial jellyfish, behaving like a single complex organism but built from a whole crowd of separate, separately reproducing individuals? What about mangroves, cuckoos, lobsters, kangaroos, hummingbirds, wandering albatrosses, and blue whales that subsist purely on plankton?

WHAT, NO ALIEN BEINGS?

One could also ask, of course, about the status of any intelligent extraterrestrials. We might naturally expect them to join us in seeking the Omega Point, and, for the purposes of this argument about probability, we might also worry about whether they are more improbable than ourselves, and so more cosmically significant. Surprisingly, however, Barrow and Tipler are sure that there are no intelligent beings except ourselves. They give some wild reasons for this, notably that such beings, if they existed, would by now have colonized space and would have already arrived here.

By this same argument, of course, these beings can prove that we do not exist either. This must be a great comfort to them, especially if they have heard of the fall-back plans which Barrow and Tipler suggest for communicating with them, should they show up after all:

The probe could construct an artefact in the solar system of the species to be contacted, an artefact so noticeable that it could not possibly be overlooked. If nothing else, the von Neumann probe could construct a 'Drink Coca-Cola' sign a thousand miles across and put it into orbit around the planet of the other species.[10]

On the whole, however, the possibility of other intelligent life is simply not considered. This dismissal is surely one more striking example of the imaginative dullness and inertness that I mentioned earlier.[11] Anthropicists are people who are simply not interested in the possibility that there might be conscious and intelligent beings that are radically unlike humans. Arguing from the currently accepted view that conditions for evolving carbon-based life similar to our own elsewhere are highly improbable, they conclude both that the possibility of such life can be ignored, and that there can be no other kind of life possessing intelligence.

More surprisingly still, though they follow Dyson in proposing to model future post-human beings rather closely on the intelligent Black Cloud in Fred Hoyle's story,[12] they also argue, quite straight-faced, that there cannot really be such intelligent clouds out there already, because they could never have evolved on their own. This, of course, is to miss the whole point of Hoyle's story. What Hoyle did so well was to sketch out, for once, a convincing alien being of a kind radically different from ourselves. The moral of his story is not that there may actually be that particular kind of being. This may well be totally impossible, and it is astonishing to find Dyson citing the story as if it were a scientific article showing that the scheme was practical, and adding Karel Capek's fantasy about Robots, *RUR*, as some kind of further confirmation.[13] The real moral of such stories is simply that we have not the slightest idea about the possibility of intelligent beings unlike ourselves, and had better not make fools of ourselves by dogmatizing about it.

SUMMARY

To sum up on this argument – if the question is really one about probability, we know of thousands of other organic and non-organic items that are just as unlikely as ourselves. Should we just add the giraffic and pandic principles and the others to our list? Or would it be better to have a single biospheric principle? The whole biosphere, including us all, is surely much more improbable even than its oddest group of inhabitants.

A biospheric principle may well have more appeal than a red-spottic one. It might do so even if astronomers find that the red spot actually requires even more particular conditions than our biosphere does. This brings out well how little concern this argument really ever had with the question of probability, and how directly it flows from quite traditional teleological thinking. It is because our own life seems to us so important that we are struck and shaken by finding it so improbable. Equal or greater improbability in matters we do not care about does not surprise us at all.

o humo-centric view [margin annotation]

THE ARGUMENT FROM THE NEED
FOR OBSERVERS

It is hard, then, to see how this argument from improbability could ever pick out the human race as the Strong Anthropic Principle demands. The whole burden of proof is thus left resting on the argument from our standing as observers, the only one that might give humans unique cosmic status. If we are indeed the only observers in the universe, and if the universe is not real until it is observed, then 'the universe must have those properties which allow life to develop within it at some time in its history', as the Strong Anthropic Principle says.

John Wheeler's position is indeed that the universe needs us to make it real in this way. Thus, in the passage we have already glanced at:

> Beginning with the big bang, the universe expands and cools. After eons of dynamic development, it gives rise to observership. Acts of observer-participancy – via the mechanism of the delayed-choice experiment – in turn give tangible 'reality' to the universe not only now but back to the beginning.[14]

There is no explanation of how it has been possible for these eons of dynamic development to take place without tangible reality. Wheeler, however, goes on to explain that making these observations is an elementary act of creation. It reaches into the present from billions of years in the past . . . Useful as it is under everyday circumstances to say that the world exists 'out there', independent of us, that view can no longer be upheld. There is a strange sense in which this is a 'participatory universe'.

Or, as Barrow and Tipler claim with satisfaction:

Modern quantum physics gives the observer a status that differs radically from the passive role endowed by classical physics. The various interpretations of quantum mechanical measurement . . . reveal a quite distinct Anthropic perspective from the quasi-teleological forms involving the enumeration of coincidences which we described in detail in the preceding two chapters.[15]

We had better take this story in stages. First:

The limits of causal reasoning

1 How much weight could this argument from observership carry if it did succeed?

Not much. At best, it refers only to the past. It reasons causally, claiming that the universe would not have become real without observers, just as a tornado would not have occurred without suitable weather conditions. Observers are thus placed among the other causes that must be assumed to have operated if the (real) universe was to get into its current state, though they are – somewhat surprisingly – given a unique licence to act retrospectively.

Causal arguments can, however, establish only the Weak Anthropic Principle. They only say how the world has developed. They do not give humans any special importance in it now or in the future. To have been useful in this way might please us, but it could make our status no grander now than that of the other causes involved – *unless*, of course, it is also assumed that we have been fulfilling some cosmic purpose for which we shall still be needed. Without that purpose, causal arguments do not help the Final Anthropic Principle. They cannot show that the universe is not a tornado but a soufflé which will be kept cooking till we complete our destiny. They give us no reason at all for colonizing space or aiming for the Omega Point or expecting to become immortal.

Who counts as an observer?

2 Are we the only observers in the universe?

Extraterrestrials having been excluded, do humans alone have this remarkable power to confer reality or can animals do it too? And again, does any human observer do it, or is it only done by qualified physicists, or perhaps by their apparatus?

Anthropicists ignore animals, and say little about non-scientific human observation. The question about animals is familiar in the idealist tradition that lies behind these theories, but did not seem important when people were believed to have been present from the Creation. Now, however, we know that humans are late-comers. If this remarkable retrospective trick of making-the-world-have-been-real-all-along can be performed simply by perception, then it was done by the first sentient animals, long before there were people. The principle should then be called something like the Animalic or Sentic Principle.

Why is this animal perception not enough to confer reality? The reason seems to be that humans are being drafted into the place of God, both as causal agents in creation and as an explanation – a guiding intelligence that can make the plan of the universe comprehensible. Animals are, I think, not mentioned because they are not seen as intelligent enough to fill this role. Anthropicists find it natural to focus their brand of idealism, not just on human perception, but on the special kind of observation that is involved in certain experiments in quantum mechanics. Indeed, towards the end of the book it emerges that perhaps the right kind of observation can only be performed by the Ultimate Observer, who will not exist until the Omega Point. 'Not until "then" is the Universe actualised.'[16] We need, therefore, to ask next:

Will idealism help?

3 What kind of idealism is intended here?

Apparently, it is quite close to the traditional kind. John Wheeler, seeing Berkeley heaving up on his horizon, hails him as a colleague. He writes:

> How does quantum mechanics differ from what George Berkeley told us two centuries ago, '"Esse est percipi" to be is to be perceived'? Does the tree not exist in the forest unless there is someone there to see it? Do Bohr's conclusions about the role of the observer differ from those of Berkeley? Yes, and in an important way. Bohr deals with the individual quantum process. Berkeley – like all of us under everyday circumstances – deals with multiple quantum processes.[17]

This clearly cannot mean that Berkeley is wrong – that there actually *is* an outside world genuinely independent of us, but a world composed of mind-dependent particles. You cannot make an objective cake out of subjective flour, an independent universe out of particles shaped by the mind. Wheeler has to mean that both Bohr and Berkeley are right to think that reality depends on our minds, though he adds that Berkeley's view matters less for science because it deals with the macroscopic level – with things like trees, where quantum phenomena cancel out and 'can hardly be said to influence the event observed'.[18] ('Hardly' is good. What if they do?)

In much of his article, Wheeler commits himself deeply to this radical idealist position. For instance, he cites a Jewish legend telling how God acknowledged his debt to Abraham for celebrating him, and claims that we can similarly say to the universe, 'You, great system, are made of phenomena, and every phenomenon rests on an act of observation. You could never even exist without elementary acts of registration such as mine.'

Consistently with this, Wheeler uses human observation as the cause of the whole creation, the factor that has made it possible for the universe to arise out of nothing. He proposes 'the observer-participancy of quantum theory as the mechanism for the universe to come into being' and hazards that 'events of observer-participancy' are 'the sole blocks for building the laws of physics – and space and time themselves'.

These events have, accordingly, produced and shaped the universe, which is thus 'a self-excited circuit'. (Actually, self-excited electrical circuits do not start themselves off; they merely maintain themselves by feedback. Wheeler's use of this term is unquestionably naughty.) He ends by claiming that this idea of observation as the source of the universe is a revolution in physics comparable with Einstein's, and must now be completed on the same scale. 'We have to move the imposing structure of science over onto the foundation of acts of observer-participancy.' Thus, responding no doubt to the many considerations which now give good reason for demoting physics from its former throne, he finally claims that

> physics is a magic window ... Its scope is immensely greater than we once realised. We are no longer satisfied with insights only into particles, or fields of force, or geometry, or even space and time. Today we demand of physics some understanding of existence itself.[19]

All these grand structures are built on idealism. At the same time, with an offhanded inconsistency equal to Dyson's, Wheeler drops this way of thinking whenever it suits him. He denies that observation has anything to do with consciousness, accepting Bohr's position that what actually does the observing is not human minds but the 'experimental device – grain of silver bromide, Geiger counter'[20] that is used to do the measuring.

If this really is all that is meant, then the vast cosmic claims collapse into absurdity. Measuring devices, if they are really detached from human intentions, are just physical objects. On their own, in a world where no mind uses or understands them, grains of silver bromide can no doubt exist and respond to photons. But they do not then *measure* or *register* or *record* anything at all. These words only have a sense when they describe acts carried out by enquirers. The grains could indeed still be affected causally by quantum events. But then so would other physical things, for instance the particles surrounding those events. None of these effects would have any meaning, any significance, any importance. None could possibly be credited with exciting roles in creating the cosmos.

Physics, in fact, cannot extend itself in the way Wheeler wants, nor does it need to. Its immense successes have been achieved by carefully confining its scope to the topics that suit its methods – namely, what are rightly called physical objects. The ways in which these objects are observed are indeed important for it, and its recent attention to them is essential for its work. But this cannot extend its scope to the study of 'existence itself' nor of the human subjects who set up the experiments. The study of these human subjects is something for which physics has carefully disqualified itself, and which no more belongs to it than the study of language does. For examining subjects rather than objects, quite different methods exist and must be used. They are methods that involve little if any mathematics. They centre on humility, self-awareness and attention to the details of life.

No doubt for this reason, many physicists have now taken to defining the word *observe* in these contexts so as to exclude any suggestion of consciousness. It is largely used in discussions about quantum mechanics in this merely causal sense, to mean simply that the instruments *are affected by* the particles under enquiry. Indeed, other physical objects which are affected by the particles are now said to 'observe' them too. This is a technical sense of an everyday word, which of course is quite legitimate when it is well understood. But it raises the usual trouble that affects such technical uses. Elements from the common use can cling to it, and can give, as they do here, an impression of strange powers which is really alien to the scientific doctrines being stated.

THE NEED FOR A FRAMEWORK

Wheeler's flirtation with idealism is remarkable. First, the philosophical price-tag of joining Berkeley at all – even fitfully – is heavy, and second, it surely will not buy him what he needs.

Full-scale idealism of Berkeley's kind is defensible, but it has far-reaching and unfashionable consequences. Traditional idealism centres on some unifying entity such as God, or a world-soul, or (as in Buddhism) a community of souls. These background entities are needed to supply the order and continuity that is so essential for all our thought, especially for science, once the physical world is deemed not to be independently real. Berkeley did not mean that physical things just vanished when unobserved, or that they were in any way formed by human minds. They existed timelessly as a system of ideas in the mind of God, who communicated his reasonable thoughts directly to humans through perception.

Without this kind of metaphysical framework, idealism

mutates to some more parsimonious, sceptical and puzzling form, where the grounds of continuity are problematic. Most naturally, it becomes Hume's kind of Phenomenalism, which says that experience itself is all there is. There are then neither lasting physical things perceived nor lasting minds perceiving them, but just what we now think of as the interface between them – a formless flow of perceptions or sensations, a vast, impersonal flood without a river-bed, without a source, without any known direction. In this flow, transient minds and transient bodies sometimes seem to form, but no real explanation of their presence is possible. When we claim to detect order among them, we are just constructing frameworks for our own convenience. (There are some very awkward problems here, notably one about cosmic history before sentient life evolved. But they need not worry us now.)

Obviously, neither of these ways of thinking has any room for the suggestion that human observation makes the physical world real. No such change is possible. For phenomenalism, reality belongs only and unchangeably to sensation itself. For the older systems, the hidden real thing that lies behind what we perceive and speaks through it, is spirit, not matter. In them, the physical world is and must remain simply a veil for spirit. It may be called a complete illusion, or merely, as with Plato, *less real* than spirit, an incomplete and obscure expression of it. But it cannot possibly be made any more real by being observed.

DEGREES OF REALITY

The idea that reality can have degrees, can be more or less fully present, is important in idealist thinking. Wheeler seems to invoke it too. Asking whether unobserved photons (infinitely more numerous than observed ones) are unreal, he replies,

> Of course not, but *their 'reality' is of a paler and more theoretic hue*. The vision of the Universe that is so vivid in our minds is framed by a few iron posts of true observation . . . Most of the walls and towers of the vision are of papier-maché, plastered in between those posts by a an immense labour of imagination and theory.[21] (Emphasis mine)

Similarly he talks about the 'tangible reality' conferred on the whole universe by observation. He explicitly extends that view to all ordinary perception as well. But this of course destroys any last hope there might be of picking out quantum observations as special, significant acts of creation. If Wheeler means the idea of different realities to bear any weight – if he is saying more than just that our perceptions are patchy – then he is giving the phenomenalist account, which applies equally to all experience. He is saying that all we believe about the outside world is just a construction we form for our own convenience. The paler 'reality' conferred on the unobserved world is then a mere reflection, a derivative of that belonging to all our perception – not, of course, only to observations made in quantum mechanics. The idea of an independent real world has vanished altogether.

18

QUANTUM QUANDARIES

PROBLEMS OF FOCUS

Wheeler's confusions are not private to himself. Habitually bold in expressing unconventional thoughts, he does us great service by displaying plainly and fully imagery which in other people's thinking is usually half-developed and hidden in the background. Once these ideas are stated, however, it does become important to place them, to think what is right and wrong about them.

This kind of ambivalence about idealism is widespread, only it is usually much less conscious. It is essential for the *folie de grandeur* involved in G. Wald's claim, which Barrow and Tipler quote, that 'a physicist is an atom's way of knowing about atoms'[1] and that, more generally, physicists are the universe's way of knowing about the universe. It should surely be obvious that, if the universe is the kind of thing that is capable of knowing or wanting to know anything, it can do this on its own and does not need kindly help from physicists.

This confused half-intended idealism is one more symptom of the split in current attitudes to science. Prudent official modesty, designed to ward off sceptical attacks, clashes with the confidence, the bold vision, which helps people to carry on scientific work. Quantum mechanics is a special focus for this split – a kind of San Andreas Fault – because in it the gap between successful practice and floundering theory is so glaring. Its equations are constantly being applied to the world with admirable and fertile results. It has made possible the invention of lasers, electron microscopes, transistors, superconductors and nuclear power. To an extraordinary extent, it works. But so far nobody has managed to explain coherently in words just what the happenings described by these equations amount to.

In this book I cannot of course attempt to deal with that problem, nor the related paradoxes arising about things like Heisenberg's uncertainty principle and the wave properties of matter. We can consider here only the impact of such puzzles on the way scientists and others regard science. We are noting the symbolism of the debate, the forms in which new technical thinking affects everyday imagery, the emotional and imaginative difficulties the puzzle has posed, and the imaginative defences that have been raised against them.

The change from the epoch when clear explanation on the single mechanical model was expected has been rapid and dramatic. As Ivan Tolstoy puts it: *[handwritten: AGAIN W/ THE PASSIVE VOICE! ENOUGH!]*

> In the first forty years of the twentieth century, our vision of the physical world changed radically and irretrievably. Atoms could behave like solid matter or like waves, they were made of particles with strange top-like properties, with nuclei which could disintegrate spontaneously, and, perhaps, set up chains of disintegration themselves ... For many, the most interesting implication of all this new knowledge was, and still is, philosophical ... We have understood that our intuitive ideas of what is possible and what is not – our common sense – are a result of the conditioning of our minds by sense-experiences ... *We have had to change our ideas of what understanding consists in.* As Bohr said, 'When it comes to atoms, language can only be used as in poetry. The poet, too, is not nearly so concerned with describing facts as with creating images.' The same is true of cosmological models, curved spaces and exploding universe. *Images and analogies are the key* ... Not you, not I, not Einstein could interpret the universe in terms wholly related to our senses. Not that it is incomprehensible, no. But we must learn to ignore our preconceptions concerning space, time and matter, abandon the use of everyday language and resort to metaphor. We must try to think like poets.[2] (Emphases mine)

POETRY, HISTORY AND OTHER SKILLS

The need for some such change is now quite widely admitted in principle, but people find great difficulty in envisaging it. Scientists who have not been trained to recognize systematic thinking anywhere but in specialized science do not easily understand Bohr's and Tolstoy's advice. They have been brought up to think of poetry as something casual, amateur, emotional and formless. The idea that poets 'think', that poetry is itself a rigorous, highly disciplined art is quite foreign to them.

Bohr's advice is indeed often quoted, but it is most commonly used to justify disordered thought, the kind of free-for-all that we have encountered in this book. Apart from that, scientists usually take it as just a pious hope that we will be lucky and get inspiration from somewhere. The idea that what is needed is to attend more sharply to the words themselves – to understand better the meaning of the analogies underlying even the most apparently literal descriptions, to study the workings of metaphor – is not on their horizon at all.

It is worth while to attend for a moment to the way in which serious non-scientific writers do handle symbolism. They do not by any means just get obsessed with a single dramatic image, leap on board it, plug it for all it is worth, and sail away wherever it takes them. That is indeed a recipe for writing bad history and poetry, which of course sometimes sells better than the other kind. It is also a recipe for producing unbalanced, misleading scientific books, such as those we have been discussing. What good writers do is quite different. They use a whole constellation of related but widely varying images, balancing them against each other and forging them into a harmony, so as eventually to convey a new and complex message.

The way in which this could be done for scientific thought and writing is perhaps best seen by looking at historical method. History cannot be written by establishing universal laws through control experiments. We cannot find the causes of the French Revolution by rerunning it with fifty different variables. Instead, as well as studying the details of what happened, we also look at partial analogies with a wide range of other revolutions, near-revolutions and other similar happenings, using these to correct each other. From these varying comparisons, we get light on the various conceptual schemes that are being used to describe the events. Steady, persistent work of this kind makes it possible to allow for bias, to refine the tools used, to see what is blocking understanding in the first place, and so gradually to make what happened more intelligible.

PROTONS AND TABLES

This is surely the kind of thing that Bohr had in mind in his famous controversy with Einstein about quantum mechanics. Bohr was certainly not opting for a single, dramatic and paradoxical account of the behaviour of particles. He was not saying anything so crude as that observation created or shaped the observed particles. He was proposing an embargo on all such talk until the conceptual scheme could be made more adequate. He was saying that these particles were objects of so strange a kind – if indeed they should be called objects at all – that they could not at present be talked about except in the context of particular observations. He was, in fact, invoking instrumentalism.

Bohr saw that talking about 'a proton' is not like talking about a table. It is using an elaborate, highly abstract intellectual system. The question is, how to adapt that system to deal with a new and awkward situation. The sort of embargo that he proposed is, of course, not something that could be used at the level of everyday life. The systems there are far too large, too complex and too widely rooted in experience. If we refused to talk about unobserved tables, we could not deal adequately with what we need to say about observed ones. We encounter tables in far too many ways for such a rule to be workable. (As we have seen, J-L.Borges wrote a story about a world where nobody does believe in anything unobserved, and tried to work out a scheme for its language. He surely demonstrated well why, for human beings like us, this could not be done.)[3]

Bohr, however, thought that, inside the well-defined province of quantum mechanics, this gap in the history of the proton could be confined within limits that would allow clear thinking to go on. Einstein, notoriously, disagreed. Determined that understanding of the familiar kind should be restored, he proposed many solutions that he hoped would re-establish it.

But he could not make them acceptable, and some of them were actually disproved by experiments.

PHENOMENA AND THINGS IN THEMSELVES

The Bohr–Einstein debate is not easy to follow today, partly because its philosophical background was far more sophisticated than is now usual.[4] Idealism was then a prevalent philosophical view, taking many subtle forms. It had gained ground in the nineteenth century, largely because of growing awareness that the forms of complex human thought – especially those of science – were not just copied from the world, but were to some extent imposed on it.

In what sense, then, can we say that the world exists independently of us? This really is not a simple question, and the idea of finding a flat yes-or-no answer to it was abandoned before the nineteenth century. Kant argued that this independent world – 'the thing in itself' must indeed be said to exist, but only in an indefinite, general sense. Every fact we can state about that world incorporates elements of our own thinking. The world that science studies cannot therefore be that thing-in-itself, though it is shaped by it in a way that we can never fully understand.[5]

Science, then, (said Kant) deals with a world of 'phenomena'. This word, which originally just meant 'appearance', might sound as if science is wholly subjective. But Kant, as he explained, did not mean by it mere sense-impressions. He meant an ordered, apparently independent, system following its own laws. This system was something which science could to some extent discover, not just invent. For all normal purposes we must treat it as 'out there'. But we must also remember that all our descriptions of it involve the forms of our own thought. We must constantly criticize and improve these, but we can never stand right outside them.

This history matters because the great physicists who first struggled with quantum mechanics – Bohr, Einstein, Heisenberg and the rest – were well read in philosophy, especially in the philosophy of Kant. They took his ideas for granted, including this quite subtle, technical sense of 'phenomenon'. It cannot be right, therefore, for Wheeler to treat them as starting much further back, with a far cruder set of alternatives. Wheeler represents Einstein as insisting, blankly and crudely, 'that the universe exists "out there" independent of all acts of observation'. Against this crude position, he says, Bohr invented an equally extreme one:

> Bohr found himself forced to introduce the word 'phenomenon'. In today's words, Bohr's point – and the central point of quantum theory – can be put in a single, simple sentence. 'No elementary phenomenon is a phenomenon until it is a registered [observed] phenomenon.'[6]

But unless we know just what a 'phenomenon' is, this is, unfortunately, not simple at all. Bohr, plainly, was talking about the use of that word, about what could properly be described as a 'phenomenon' – an object of science. He was insisting that the method of observation must be incorporated into the description. Thus, talk about protons or the like is always talk about them as observed in certain ways. Any enquiry about what they do the rest of time is empty. It would therefore, of course, also be empty to say that anything (such as observation) caused them to take on particular qualities.

Wheeler takes Bohr's remark to justify his own causal claim, to mean that, 'we (as observers) are inescapably involved in bringing about that which appears to be happening'. This has to mean that our observations cause the proton to behave as it does. Wheeler's extra words 'appears to be' might indeed seem to contradict this, but then his whole exciting conclusion would be wrecked. Such qualifications, though peppered fairly freely, seem to be no more than disconnected relics of an early training in caution.

These brief remarks on a most difficult topic are, I fear, none too clear, but I hope they may make one crucial thing plain. Bohr's Copenhagen Interpretation of quantum mechanics was meant above all to be modest and parsimonious, to say as little as possible about anything beyond what was observed. That was the main reason why it found favour and has become to some extent regarded as the 'accepted' one. Anthropicists, however, persistently misunderstand this parsimonious intention. They read Bohr's refusal to talk about what brings particles into their observed state as a positive claim that observation itself is the cause that brings them there.

It must, I think, be said that Bohr himself is partly to blame for these confusions. It is not half as easy as he seems to have hoped to say nothing about what lies behind what you observe. Some ideas about what photons are and how they act are bound to be involved in reporting the observations at all. Suspensions of illicit explanations are not themselves explanations. They are just temporary notices hung up to say that no explanation can, for the time, be given. Indeed, it is not clear that instrumentalism can ever be fully coherent outside its original idealist habitat. It is obscure how theories could ever be useful 'instruments' if they were not in some way accepted and believed.[7] The Copenhagen Interpretation is perhaps more a promise of an interpretation than the thing itself. Such promises are vacuums that present a standing temptation to fantasists who enjoy filling them.

19

CONSERVING THE SPIRIT

MECHANISTIC HOPES

We must, I fear, draw to a close, leaving many fields of anthropic daisies unpicked. It is tempting to spend time on the Many Worlds Interpretation of quantum mechanics, much odder in itself than the Copenhagen one, and figuring even more oddly in these stories. Discussing it would not, however, be so much to our present purpose because that theory, being less anthropocentric, has, despite its oddity, less myth-building power.

Again, I think we must regretfully leave aside that engineer's nightmare, the Von Neumann Probe – the self-reproducing universal-construction-machine, due to be sent to other galaxies in order to build further intelligent machines which will do the colonizing for us, while the Probe duplicates itself and repeats the process. It is certainly interesting that this mode of colonizing quietly drops the two aims that were most important to Dyson, since it neither relieves the human population problem nor provides people with adventures. It is intriguing, too, that machines which have never met a person or lived on earth, and which (if the Turing Test is indeed relied on) may not even be conscious, are expected to carry on human culture so smoothly as, in effect, to continue the species itself. Barrow and Tipler explain that to doubt this would be a piece of culpable racism:

> An advanced von Neumann probe would be an intelligent being in its own right, only made of metal rather than flesh and blood . . . The arguments one hears today against considering intelligent computers to be persons and against

giving them human rights have precise parallels in the nineteenth-century arguments against giving blacks and women full human rights ... If [the current] anti-racist trend continues and occurs in the cultures of all civilised beings, von Neumann probes would be recognized as intelligent fellow-beings, beings which are heirs to the civilization of the naturally evolved species that invented them ... If a naturally evolved species never has machine descendants, its civilization will eventually die out. A civilization with machine descendants could continue indefinitely.[1]

There, apparently, speak two people who have never known a car to break down. A brief spasm of doubt does strike them about the possibility that some of these new communities might disagree among themselves – might, in fact, start fighting. This they counter by predicting that intelligent beings are surely more likely to respect each others' freedom ... They then propose a few obviously disastrous ways of dealing with such disagreements, and move hastily on to quarrel with Carl Sagan about the presence of extraterrestrials.

OTHER-WORLDLY PRECEDENTS

What, however, has been the point of the whole thing? What had Bernal in mind when he wrote that 'the new life which conserves none of the substance and all the spirit of the old would take its place and continue its development'? Is there indeed an invaluable spirit here, one which can be conserved in so surprising a way? Beyond the crude motives of desire for power and fear of death which we have already noted, what ideal are these visionaries really pursuing?

On the face of it, their quest seems to have a remarkable amount in common with traditional religious movements:

1 They are in a very obvious sense 'other-worldly', pointing us away from the earth, the flesh and our familiar worldly pleasures.
2 In doing this they evidently posit a 'soul' of some kind essentially separate from the body. Their position about this seems to be nearer to Plato and Descartes than to Christianity or Buddhism, because they conceive this soul as

primarily intellectual. But that is surely a minor difference in comparison with the startling fact of putting it there at all.

3 They seem to think it appropriate for this soul to dismiss its present body with contempt as unimportant to it. In this contempt for the flesh and for other earthly things, they go, in fact, considerably beyond the religious positions just mentioned, all of which do something to balance this world-denying spirit by a recognition of the legitimate claims of the flesh and of our continuity with the rest of nature.

Plato wrote the *Timaeus* and the *Symposium* as well as the *Phaedo*. Buddhists are related to all other living creatures by reincarnation. Aquinas brought in Aristotelian doctrines to balance the world-denying elements in Christianity. Indeed, the only religion in our tradition which has completely renounced physical matter as evil and alien to the spirit is Manichaeism. That, however, can hardly be the position of these physicists, if only because they take the souls, when released from their earthly prison, to find refuge and new bodies in outer space. If matter is vile, it is still just as vile however many light-years away it is, and however gaseous you make it.

THE PURSUIT OF THE LITERAL

It is, however, certainly very striking how close the anthropicists' symbolism is to that used by religious sages. Light, knowledge, the heavens, immortality, the 'upward' movement, and the increase of power, all sound very similar. *But the use made of these symbols is quite different, indeed contrary.* That use has now become wholly literal, offering salvation by technical fix. Trust is placed wholly in miraculous machinery. There is no suggestion of changes in the inner life that might underlie these miracles and explain their point.

Moreover and relatedly, the grand change aimed at lies wholly in the remote future. It does not involve any present conversion, any immediate change in moral attitude, as Utopias mostly do. It is as if mystical 'enlightenment' would be reached in the future merely by turning on an electric light, as if the spirit would rise to its true destiny merely by getting into a lift. Centrally, perhaps, the metaphor is that of distance, of the need for something at present quite unknown. In Dyson's picture, the prospect of very distant glory is the only thing needed to give life a meaning. But if life in the present really has no meaning, how could these remote prospects ever provide it?

This flight from the near to the remote is the really odd thing, as Francis Thompson saw:

> Does the fish soar to find the ocean?
>   The eagle plunge to find the air –
> That we ask of the stars in motion
>   If they have rumour of thee there?
>
> Not where the wheeling systems darken
>   And our benumbed conceiving soars –
> The drift of pinions, would we hearken,
>   Beats at our own clay-shuttered doors.
>
> The angels keep their ancient places –
>   Turn but a stone and start a wing!
> 'Tis ye, 'tis your estranged faces
>   That miss the many-splendoured thing . . .
>
> But (when so sad thou canst not sadder)
>   Cry, and upon thy so sore loss,
> Shall shine the traffic of Jacob's Ladder
>   Pitched between Heaven and Charing Cross.[2]

At their best, the great religious traditions have never made the mistake of glorifying mere distance, any more than the humanistic tradition did – until lately. For both these great traditions, the meaning of life is now, is here. If it wasn't here and now, the future could not supply it. When religious sages speak of eternity, they don't just mean a very long time which doesn't happen to have an end. They mean timelessness, which is chiefly found in the present moment.

If, however, we ask what has been, up till now, the special excellence of the humanistic tradition, we might well say that it has simply grasped this point more clearly. Humanists who do not believe in God or a future life have been in a stronger position to insist on the urgency of making things better at once, in this one. If this is the only life that anybody has, then the fact that many people must spend it in such misery becomes more obviously and inexcusably scandalous. Salvation is needed now; it can't be put off to some vaguely planned future state. But if these dreams are taken seriously and accepted as answers to the problem of salvation, that strong point in humanism goes right out of the window. A.E. Housman got that right:

> From far, from eve and morning
> And yon twelve-winded sky,
> The stuff of life to knit me
> Blew hither – here am I.
>
> Now – for a breath I tarry,
> Nor yet disperse apart –
> Take my hand quick and tell me
> What have you in your heart?
>
> Speak now and I will answer
> How can I help you, say?
> Ere to the wind's twelve quarters
> I take my endless way.[3]

– Is it plain now why I have asked you to take the trouble of examining these strange dreams? The notion they convey that our natural, earthly life can be despised is not just meaningless; it is disastrous. It is not just a scheme for what might some time be done in outer space. It promotes, here and now, a distorted idea of what a human being essentially is. Its suggestion that our biosphere is merely so much waste matter and the human body, at best, a rather unsatisfactory ship in which the intellect has to sail, expresses an unrealistic, mindless exaltation of that intellect – narrowly conceived as searching for facts – and a corresponding contempt for natural feeling.

If we try to bring together in our minds this euphoric estimate of what the human intellect is and can achieve with a real sense of our current difficulties and dangers, the result can only be a most painful confusion and vertigo. As commonly happens, therefore, people do not bring the two together; they carefully keep them apart. They retreat into one of the usual techniques for cobbling up cognitive dissonance. These notoriously tend to involve the public chanting of songs of triumph in order to strengthen whatever tribal bonds we may have formed and to distract us from the upsetting discrepancies before us.

On the present topic, the songs of triumph lie ready in great numbers. For the last three centuries, able people have been celebrating the astonishing achievements of the human intellect and the human will. Like Ozymandias, they have continually exclaimed, 'Look on my works, ye Mighty, and despair.'[4] The dominant world picture in our culture has been one of steady linear progress brought about by that will and intellect, an improvement booked to continue indefinitely.

This remarkably high opinion of ourselves is not needed to support human life. Even in our own culture it is a recent growth, and nobody else has carried it anywhere near so far. We have, however, grown in some degree addicted to it; it is hard to wean ourselves from it. Of course there have been protests against it. But for a long time these protests were seen as voices from the past – a past which (on this assumption) was by definition bound to be mistaken. Conceit, by contrast, was always the voice of the Future.

The quasi-scientific dreams I have been discussing are surely a rather frantic and defensive last outbreak of these celebrations. Most of us have begun to see that the party is over. The planet is in deep trouble; we had better concentrate on bailing it out. At this point, to keep up one's spirits by further orgies of self-congratulation may be a natural reaction, but it is a dead end. Paranoia, if further encouraged, is liable finally to undermine all wish to get back in touch with reality. The discrepancy between image and fact is growing too wide to be tolerated. For the general sanity, we need all the help we can get from our scientists in reaching a more realistic attitude to the physical world we live in.

NOTES

1 SALVATION AND THE ACADEMICS

1 I cannot embark on an Honours List here, but outstanding among them is surely Stephen Jay Gould.
2 C.H. Waddington, *The Scientific Attitude* (West Drayton, Penguin, 1941), p. 81.
3 J.D. Bernal, *The Social Function of Science* (London, Routledge & Kegan Paul, 1939), concluding paragraph.
4 K.R. Popper, *Objective Knowledge* (Oxford, Clarendon Press, 1972), p. 237.
5 *The Scientific Attitude*, p. 170.
6 Richard Dawkins, *The Selfish Gene* (London, Granada, Paladin, 1978), p. 1.
7 Stephen W. Hawking, *A Brief History of Time* (London, Bantam Press, 1988), p. 13.
8 ibid., p. 175.
9 See a most illuminating discussion in Graham Richards, *On Psychological Language* (London, Routledge, 1989), chapter 1.
10 Aristotle, *Nicomachean Ethics*, Book I, chapters 1–8, Book X, chapters 7–8.
11 See a helpful discussion in the opening sections of *On Intention* by Elizabeth Anscombe (Ithaca, Cornell University Press, 1963).
12 Its invention is usually ascribed to C.S. Pittendrigh, 'Adaptation, natural selection and behaviour', in *Behaviour and Evolution*, ed. A. Roe and G.G. Simpson (New Haven, Yale University Press, 1964), pp. 390–416.
13 See Max Hocutt's article 'Aristotle's four becauses' in *Philosophy*, 49 (1974). It is interesting that Pittendrigh, in the article just mentioned, explicitly makes this very mistake, saying that he wants to avoid 'carrying a commitment to Aristotelian teleology *as an efficient causal principle*'. Aristotle carefully distinguished the 'efficient cause' – the precipitating event – from the point or purpose, which is the teleological element.
   I have discussed this pervasiveness of teleological thinking

225

briefly in my *Beast and Man* (London, Methuen, 1980), pp. 72–5, and more fully in 'Teleological theories of morality', in *An Encyclopaedia of Philosophy*, ed. G.H.R. Parkinson (London, Routledge, 1988).

14 Stephen Clark's book *The Mysteries of Religion* (Oxford, Blackwell, 1986) is very helpful in breaking up the stereotypes that underlie this notion. See also a very interesting article by John Bowker, 'Did God create this universe?' in Arthur Peacocke (ed.), *The Sciences and Theology in the Twentieth Century* (Notre Dame, Indiana, University of Notre Dame Press, 1981).

15 For this whole dispute, see the opening sections of *Quantum Theory and Measurement*, ed. J.A. Wheeler and W.H. Zurek (Princeton, Princeton University Press, 1983). We will return to this topic in chapters 17 and 18.

16 C.S. Lewis, *Christian Reflections* (Glasgow, Collins, Fount, 1967), p. 89.

17 *Evolution as a Religion* (London, Methuen 1985).

2 PROPHECIES, MARXIST AND ANTHROPIC

1 John D. Barrow and Frank J. Tipler, *The Anthropic Cosmological Principle* (Oxford and New York, Oxford University Press, 1986), concluding paragraph.

2 See chapter 3, p. 9.

3 J.D. Bernal, *The World, the Flesh and the Devil* (London, Cape, 1929), pp. 35–6.

4 Freeman J. Dyson, 'Time without end: physics and biology in an open universe', *Review of Modern Physics*, vol. 51, no. 3 (July 1979), p. 447; p. 448.

5 *The Anthropic Cosmological Principle*, pp. 674–5.

6 'Time without end', pp. 447 and 459–60.

7 Paul Davies, *Superforce: The Search for a Grand, Unified Theory of Nature* (Unwin Paperbacks, 1984), pp. 167–8.

8 *The World, the Flesh and the Devil*, pp. 56, 51.

9 'Time without end', p. 453.

10 *The Anthropic Cosmological Principle*, pp. 154, 523, 659.

11 Steven Frautschi, 'Entropy in an expanding universe', *Science*, vol. 217, no. 4560, (August 1982), p. 593; pp. 598–9.

12 ibid., p. 599.

13 ibid., p. 658.

14 ibid., p. 15

15 *Critique of Pure Reason*, Introduction, section VI, 'The general problem of pure reason'.

16 *The World, the Flesh and the Devil*, p. 83.

17 ibid., p. 96.

18 *Possible Worlds* (London, Chatto & Windus, 1927) p. 287.

19 ibid., pp. 311–12.

20 In *Daedalus, Or The Future of Science* (London, Kegan Paul, 1923).

21 'Law without law' by J.A. Wheeler, in *Quantum Mechanics and Measurement*, ed. J.A. Wheeler and W.H. Zurek (Princeton, Princeton University Press, 1983), p. 209.
22 *The Anthropic Cosmological Principle*, p. 21.
23 ibid., p. 22.
24 ibid., p. 23.
25 Heinz Pagels, *Perfect Symmetry* (London, Michael Joseph, 1985), pp. 358–9.

3 MINIMALISM DOES NOT WORK

1 Ilya Prigogine and Isabelle Stengers, *Order out of Chaos: Man's New Dialogue with Nature* (London, Collins, Fontana, 1984), p. 9.
2 ibid., p. 14.
3 ibid., p. 7.
4 ibid., p. 54.
5 ibid.
6 Jacques Monod, *Chance and Necessity*, trans. Austryn Wainhouse (London and Glasgow, Collins, Fount, 1977), p. 163.
7 David Hume, *Treatise of Human Nature*, Book III, part i, section 1; p. 466 in Clarendon Press edition (2nd edn, Oxford, 1978), ed. L.A. Selby-Bigge and revised P.H. Nidditch.
8 *Order out of Chaos*, pp. 8–9.
9 ibid., p. 8.
10 *Chance and Necessity*, p. 137.
11 I discussed this drama with a slightly different bearing in chapters 9 and 10 of *Evolution as a Religion* (London, Methuen, 1986). I am sorry for a certain repetitiousness here, but the point is important and does, I think, have to be explained for the present purpose.

4 THE FASCINATION OF CHANCE

1 Peter Atkins, *The Creation* (Oxford and San Francisco, W.H. Freeman, 1987), p. 51.
2 ibid., pp. 99–105.
3 ibid., pp. 29–31, 37.
4 Jacques Monod, *Chance and Necessity*, trans. Austryn Wainhouse (London and Glasgow, Collins, Fount, 1977), p. 160.
5 Steven Weinberg, *The First Three Minutes* (London, André Deutsch, 1977), p. 155.
6 *Chance and Necessity*, p. 113.
7 *The Creation*, pp. 35 and 39.
8 ibid., p. 113.
9 J.-P. Sartre, *Nausea*, trans. Robert Baldick (Harmondsworth, Penguin, 1965), pp. 182–8.

5 THE FUNCTION OF FAITH

1 A much more natural, uncontentious way of relating them is recommended by John Polkinghorne in his excellent little book *One World: The Interaction of Science and Theology* (London, SPCK, 1986), and by Hanbury Brown in *The Wisdom of Science: Its Relevance to Culture and Religion* (Cambridge, Cambridge University Press, 1986).
2 Robert Jay Lifton, *The Broken Connexion: On Death and the Continuity of Life* (New York, Basic Books, 1979), pp. 286–7.
3 Good correctives to the stereotyped idea of a simple warfare may be found in much recent historical work, notably in *The Post-Darwinian Controversies* by James Moore (Cambridge, Cambridge University Press, 1979). *Darwin's Forgotten Defenders* by David N. Livingstone (Edinburgh, Scottish Academic Press, 1987) gives a useful account of strong support for the new theories from within the Victorian evangelical movement, often thought of as solidly Creationist.
4 William James, *The Will to Believe* (New York, Dover Publications, 1956), p. 29.
5 James Lovelock, 'Stand up for Gaia'. Schumacher Lecture 1988. Reprinted by *Resurgence* (Ford House, Hartland, Bideford, Devon).
6 On the problems endemic to this quest for objectivity, see Thomas Nagel, *The View from Nowhere* (Oxford and New York, Oxford University Press, 1986).

6 ENLIGHTENMENT AND INFORMATION

1 G.E. Moore, *Principia Ethica* (Cambridge, Cambridge University Press, 1948), pp. 188–9.
2 Printed in Kant's *Foundations of the Metaphysics of Morals*, trans. Lewis White Beck (Indianapolis and New York, Bobbs-Merrill Library of Liberal Arts, 1959), p. 85.
3 John D. Barrow and Frank J. Tipler, *The Anthropic Cosmological Principle* (Oxford and New York, Oxford University Press, 1986), p. 677.
4 From *The Lancet*, vol. 1 (1870) p. 188. Quoted by Brian Easlea, *Witch-Hunting, Magic and the New Philosophy: An Introduction to the Debates of the Scientific Revolution* (Brighton, Harvester Press, 1980), p. 252
5 J.S. Mill, *On Liberty*, chapter 3. In *Utilitarianism Including Mill's 'On Liberty' & Essay on Bentham*, ed. Mary Warnock (London and Glasgow, Collins, Fontana, 1962), p. 188.
6 ibid., pp. 162–3.
7 Aristotle, *Nicomachean Ethics*, Book X, chapter 7 and Book VI, chapter 6.
8 Salman Rushdie, 'Is nothing sacred?', Herbert Read Memorial Lecture (Cambridge, reprinted Cambridge Granta, 1990), p. 8.

7 PUTTING NATURE IN HER PLACE

1 Descartes, *Le Monde*, in F. Alqui (ed.), *Oeuvres philosophiques de Descartes* (Paris, Garnier Frères, 1973), vol. 1, p. 349. Quoted by Brian Easlea, *Science and Sexual Oppression* (London, Weidenfeld & Nicolson, 1981), p. 72.
2 *The Works of the Honourable Robert Boyle*, ed. T. Birch (London, 1722), vol. 5, p. 532. Quoted by Brian Easlea, *Witch-Hunting, Magic and the New Philosophy, An Introduction to the Debates of the Scientific Revolution* (Brighton, Harvester Press, 1980), p. 138.
3 *Works*, vol. 5, p. 165. Easlea, *Witch-Hunting*, p. 139.
4 *Descartes on Method, Optics, Geometry and Meteorology*, trans. P.J. Olscamp (Bobbs-Merrill, 1965), p. 361. Easlea, *Witch-Hunting*, p. 117.
5 Peter Atkins, *The Creation* (Oxford and San Francisco, W.H. Freeman, 1987), p. 17.
6 B. Farrington, *The Philosophy of Francis Bacon* (Liverpool University Press, 1970), pp. 93, 92, 96, 92, 62.
7 Adam Sedgwick, 'Vestiges of the natural history of creation', *Edinburgh Review*, vol. 82 (1845), p. 16. Quoted by Easlea, *Science and Sexual Oppression*, p. 103.
8 Sedgwick, 'Vestiges', p. 23.
9 Francis Bacon, *Novum Organum*, vol. 4, p. 109. Quoted by Easlea, *Witch-Hunting*, p. 128.
10 *Experimental Philosophy* (1664). Introduction by M. Boas Hall (London, Johnson Reprint, 1966), p. 192.
11 'Epistle to the reader', penultimate paragraph.
12 C.H. Waddington, *The Scientific Attitude* (West Drayton, Penguin, 1941), p. 80.
13 William Gilbert, *De Magnete* (New York, Dover Publications, 1968), p. 309. Quoted by Easlea, *Witch-Hunting*, p. 91.
14 Quoted in R. Dugas and P. Costable 'The birth of a new science, mechanics', in R. Taton (ed.), *The Beginning of Modern Science* (London, Thames & Hudson, 1964), p. 265.
15 *New Philosophy of Our Sublunary World*, quoted in P. Duhem, *The Aim and Structure of Physical Theory*, trans. P.P. Wiener (1914, London, Athenaeum, 1962), p. 230.
16 Kepler, 'Astronomia Nova', trans. A.R. Hall, in *Nature and Nature's Laws*, ed. M. Boas (New York, Harper, 1970), p. 73.

8 THE REMARKABLE MASCULINE BIRTH OF TIME

1 Descartes, *The Principles of Philosophy*, part 4, para. 187.
2 Descartes, *The Principles of Philosophy*, in F. Alquié (ed.), *Oeuvres philosophiques de Descartes* (Paris, Garnier Frères, 1973), p. 502, note.
3 Descartes, *Primae Cogitationes circa Generationem Animalium* (1701)

quoted in J. Roger, *Les Sciences de la vie dans la pensée française du XVIIIme siècle* (Paris, Armand Collin, 1963), p. 146.

4 Peter Atkins, *The Creation* (Oxford and San Francisco, W.H. Freeman, 1987), p. 53.

5 Brian Easlea, *Science and Sexual Oppression* (London, Weidenfield & Nicolson, 1981), p. 73.

6 See Easlea, *Science and Sexual Oppression*, index.

7 *The Creation*, p. 127.

8 Stephen W. Hawking, *A Brief History of Time* (London, Bantam Press, 1988), concluding passage.

9 Robert Hooke, 'A discourse on the nature of comets' (1682), in R. Walker (ed.), *The Posthumous Works of Robert Hooke* (London, 1705), pp. 171–2. See Easlea, *Science and Sexual Oppression*, p. 85.

9 UNEXPECTED DIFFICULTIES OF DEICIDE

1 Plato, *Republic*, Book X, 611–12.

2 This whole style of thought received immense support from two best-selling books, A.J. Ayer's *Language, Truth and Logic* (London, Gollancz, 1936) and Bertrand Russell's *History of Western Philosophy* (London, Allen & Unwin, 1946). Both convinced a wide public that it would be a waste of time to pay any more attention to philosophy, a message which Russell at least may not altogether have meant to convey.

3 David Hume, *Treatise of Human Nature*, Book I, part iv, section 1, especially p. 185 in Clarendon Press edition (2nd edn, Oxford, 1976), ed. L.A. Selby-Bigge and revised P.H. Nidditch.

4 In *Charles Darwin and Thomas Henry Huxley: Autobiographies*, ed. Gavin de Beer (London, Oxford University Press, 1974), p. 54.

5 ibid., pp. 50, 55–6.

6 Thomas Nagel, *The View from Nowhere* (London, Oxford University Press 1986), pp. 78–9.

10 THE UNINHABITABLE VACUUM

1 In 'Empiricism in the philosophy of science', in *Images of Science* ed. P. Churchland and C.A. Hooker (Chicago, University of Chicago Press, 1985), p. 258. Quoted by Anthony O'Hear in *The Element of Fire: Science, Art and the Human World* (London, Routledge, 1988), p. 12.

2 'The empty universe', in *Present Concerns* (London and Glasgow, Collins, Fount, 1986), pp. 81–4.

3 C. Lévi-Strauss, *The Savage Mind* (London, Weidenfeld & Nicolson, 1972), p. 220. Quoted in a very illuminating discussion of 'the anthropomorphic illusion' by Graham Richards, *On Psychological Language* (London and New York, Routledge, 1989), p. 6.

4 The extreme oddness, from an evolutionary point of view, of

supposing people to know nothing of the world they inhabit until they invent it themselves is well discussed in Konrad Lorenz's excellent little epistemological study, *Behind the Mirror*, trans. Ronald Taylor (London, Methuen, 1976).

5 William James, *The Will to Believe* (New York, Dover Reprint, 1956), pp. 26–7.

6 ibid., pp. 27–8.

7 See his paper on 'Duties towards animals and spirits', in *Lectures on Ethics*, trans. Louis Infield (London, Methuen, 1930).

8 Kant, *Critique of Practical Reason*, concluding passage.

9 *The Will to Believe*, p. 10.

11 PARSIMONY, INTEGRITY AND PURITANISM

1 S.J. Gould, *The Flamingo's Smile: Reflections in Natural History* (Harmondsworth, Penguin, 1987), p. 393.

2 *The Problems of Philosophy* (London, Oxford University Press 1946), p. 63. Anyone who is inclined to think that induction does not need faith should find this whole discussion interesting.

3 Rom Harré, *The Philosophies of Science* (Oxford and New York, Oxford University Press, 1972), p. 88. See also a very interesting discussion, stressing the limitations of instrumentalism, in *What is This Thing Called Science?* by A.F. Chalmers (Milton Keynes, Open University Press, 1978), p. 115.

4 George Berkeley, *Principles of Human Knowledge*, section 58 (London and Glasgow, Collins, Fontana, 1962), p. 93.

5 J.-L. Borges, 'Tlon, Uqbar, Orbis Tertius', in *Labyrinths* (London, Penguin, 1970), pp. 32–3,

6 In chapters 17 and 18.

12 QUESTIONS OF MOTIVATION

1 I have discussed these questions in my *Wisdom, Information, and Wonder* (London, Routledge 1989).

2 David Hume, *Treatise of Human Nature*, Book I, part iv, section 7; p. 270 in Clarendon Press edition (2nd edn, Oxford, 1976), ed. L.A. Selby-Bigge and revised P.H. Nidditch.

3 ibid., Book I, part iv, section 1; p. 183 in Clarendon Press edition.

4 I have discussed this point more fully in the Introduction to my *Heart and Mind* (London, Methuen, 1983) and also in *Beast and Man* (London, Methuen, 1980), especially chapter 11.

5 *Treatise of Human Nature*, Book I, part iv, section 2; p. 218 in Clarendon Press edition.

6 See Hume, *Enquiry Concerning the Principles of Morals*, paras 192, 221–3 and 235.

7 Peter Atkins, *The Creation* (Oxford and San Francisco, W.H. Freeman, 1987), p. vii.

8 C.H. Waddington, *The Scientific Attitude* (West Drayton, Penguin, 1941), p. 32.
9 *Treatise of Human Nature*, Book I, part iv, section 7; p. 264 in Clarendon Press edition.
10 Jacques Monod, *Chance and Necessity*, trans. Austryn Wainhouse (London and Glasgow, Collins, Fount, 1977), p. 158

13 THE HUNGER FOR SYNTHESIS

1 Jacques Monod, *Chance and Necessity*, trans. Austryn Wainhouse (London and Glasgow, Collins, Fount, 1977), p. 165
2 ibid., p. 164.
3 This cult is the topic of my book *Evolution as a Religion* (London, Methuen, 1985).

14 EVOLUTION AND THE APOTHEOSIS OF MAN

1 William Day, *Genesis on Planet Earth: The Search for Life's Beginning* (East Lansing, Michigan, House of Talos, 1979), pp. 390–2.
2 'Time without end: physics and biology in an open universe', *Review of Modern Physics* vol. 51, no. 3, (July 1979), pp. 449, 454, 450, 456.
3 ibid., p. 448.
4 ibid., p. 459.
5 ibid., pp. 456, 454.
6 ibid., pp. 453–4.
7 ibid., p. 448.
8 ibid., pp. 459–60, 458, 456.
9 Keith Oatley, *Brain Mechanisms and Mind* (New York, Dutton, 1972).
10 J.D. Bernal, *The World, the Flesh and The Devil* (London, Cape, 1929), pp. 35, 78–9.
11 ibid., p. 80.
12 ibid., p. 68. The reference is to Shaw's play *Back to Methusaleh*, where the remote future life is as anti-carnal as Bernal's, but is also anti-science.
13 ibid., pp. 41–2.
14 ibid., p. 11.
15 Paul Davies, *Superforce* (Unwin Paperbacks 1984), pp. 167–8.
16 J.B.S. Haldane, *Possible Worlds* (London, Chatto & Windus, 1927), pp. 302–3.
17 ibid., p. 310.
18 *The World, the Flesh and the Devil*, pp. 67, 89, 94, 90, 94–5.
19 ibid., pp. 80, 32, 29.
20 See Maurice Goldsmith, *Sage: A Life of J.D. Bernal* (London, Hutchinson, 1980), pp. 191–7, 206–7, 234.
21 'Time without end', p. 448. See also chapter 2, p. 2.
22 See chapter 14, p. 5

23 *The World, the Flesh and the Devil*, p. 67.

15 DYSON, ANIMISM AND THE NATURE OF MATTER

1 Thomas Wright, *An Original Theory or New Hypothesis of the Universe*, 1750, facsimile reprint with introduction by M.A. Hoskin (London, Macdonald, and New York, American Elsevier, 1971). Quoted by Freeman J. Dyson in 'Time without end: physics and biology in an open universe', *Reviews of Modern Physics*, vol. 51, no. 3 (July 1979), p. 449, and in *Disturbing the Universe* (New York, Harper & Row, 1979), pp. 145–7.
2 Ivan Tolstoy, *James Clerk Maxwell: A Biography* (Edinburgh, Canongate, 1981), pp. 29 and 160.
3 *Disturbing the Universe* pp. 249–50, summarized in 'Time without end', p. 448.
4 *Disturbing the Universe*, p. 250.
5 See a very useful discussion by Graham Richards in *On Psychological Language* (London and New York, Routledge, 1989), Introduction and chapter 1.
6 J.D. Bernal, *The World, the Flesh and the Devil* (London, Cape, 1929) p. 45.
7 *Disturbing the Universe*, p. 247.
8 ibid., pp. 248–9.
9 ibid., p. 249.
10 ibid., p. 252.
11 Paul Davies, *God and the New Physics* (Harmondsworth, Penguin, 1983), p. ix.
12 Paul Davies, *The Accidental Universe* (Cambridge, Cambridge University Press, 1982), p. 123.
13 *Disturbing the Universe*, p. 252.

16 SPACE, FREEDOM AND ROMANCE

1 *The Anthropic Cosmological Principle* (Oxford and New York, Oxford University Press, 1986), p .658.
2 *Disturbing the Universe* (New York, Harper & Row, 1979), p. 111.
3 ibid., p. 117.
4 ibid., p. 233.
5 ibid., p. 234.
6 ibid., p. 234.
7 ibid., p. 235.
8 ibid., p. 230.
9 ibid., pp. 116–17.
10 ibid., p. 233.
11 ibid., p. 117.
12 ibid., pp. 116–17.

17 THE ANTHROPIC SYNTHESIS

1 John D. Barrow and Frank R. Tipler, *The Anthropic Cosmological Principle* (Oxford and New York, Oxford University Press, 1986), p. 658.
2 ibid., p. 659.
3 ibid., p. 126.
4 ibid., pp. 30 and 674.
5 ibid., p. 505.
6 ibid., p. 288, 289.
7 ibid., p. 146.
8 S.J. Gould, *The Flamingo's Smile: Reflections in Natural History* (Harmondsworth, Penguin, 1987), p. 395.
9 S.J. Gould, *Wonderful Life: The Burgess Shale and the Making of History* (London, Hutchinson, Radius, 1990).
10 *The Anthropic Cosmological Principle*, p. 591.
11 See chapter 2, p. 1.
12 Fred Hoyle, *The Black Cloud* (New York, Harper, 1957).
13 Freeman J. Dyson, 'Time without end: physics and biology in an open universe', *Review of Modern Physics*, vol. 51, no. 3 (July 1979), pp. 453, 454.
14 J.A. Wheeler, 'Law without law', in *Quantum Theory and Measurement*, ed. J.A. Wheeler and W.H. Zurek (Princeton, Princeton University Press, 1983), pp. 209 and 194.
15 *The Anthropic Cosmological Principle*, p. 505.
16 ibid., p. 471.
17 'Law without law', p. 186.
18 ibid., p. 187, 189, 206.
19 ibid., p. 210.
20 ibid., p. 207.
21 ibid., pp. 203, 209.

18 QUANTUM QUANDARIES

1 John D. Barrow and Frank R. Tipler, *The Anthropic Cosmological Principle* (Oxford and New York, Oxford University Press, 1986), p. 510.
2 Ivan Tolstoy, *The Knowledge and the Power: Reflexions on the History of Science* (Edinburgh, Canongate, 1990), pp. 15–16.
3 See chapter 11, p. 5.
4 It is thoroughly examined in the opening sections of *Quantum Theory and Measurement*, ed. J.A. Wheeler and W.H. Zurek (Princeton, Princeton University Press, 1983).
5 This is his considered view, as expressed in the *Critique of Pure Reason*. His earlier *Prolegomena to Any Future Metaphysic* had taken a simpler, more idealist position, possibly because he despaired at that time of making the more subtle one intelligible.

6 J.A. Wheeler, 'Law without law', in *Quantum Theory and Measurement*, ed. J.A. Wheeler and W.H. Zurek (Princeton, Princeton University Press, 1983), p. 184.
7 See A.F. Chalmers, *What Is This Thing Called Science?* (Milton Keynes, Open University Press, 1978), p. 115.

19 CONSERVING THE SPIRIT

1 John D. Barrow and Frank R. Tipler, *The Anthropic Cosmological Principle*, (Oxford and New York, Oxford University Press, 1986), p. 595.
2 Francis Thompson, 'The Kingdom of God'.
3 A.E. Housman, *A Shropshire Lad*, no. XXXII.
4 P.B. Shelley, 'Ozymandias'.

INDEX OF PROPER NAMES